Understanding Data Modes

By

William McFarland GM6DX

Printed in the United Kingdom of Great Britain and Northern Ireland.

Electronic version produced in the United Kingdom of Great Britain and Northern Ireland.

First Printing, 2022

ISBN 978-1-4716-6757-2

Lulu Press

627 Davis Drive Suite 300 Morrisville,

NC 27560 United States

CONTENTS

About the author

I first got into the hobby of amateur radio by rather unusual means where I purchased UHF radios to use for two-way communication during war game simulations. At one of these events a team member got an injury which resulted in him becoming lost, which in turn prompted the purchase of these two-way radios. Whilst programming the radios I scanned across a frequency which was being used by an amateur radio repeater. I heard a few call signs being aired (now both SK) so this got me intrigued and I then conducted research into the hobby. I finally visited a local club, Stirling

District Amateur Radio Society (GM6NX) and sat my radio licence exams.

Since becoming licenced in the hobby, I myself have become an instructor putting many new hams through their licences. I have constructed many radio projects from simple Morse code oscillators to more complexed PCB designs. I have participated in several DXpeditions, wrote many articles for Practical Wireless magazine and the RSGB Radcom including my first book titled *"Mini DXpeditions For Everyone"* which is being sold by the Radio Society of Great Britain.

I tend not to get hung up on amateur radio politics and I will operate on every band that I have an antenna or transceiver for. I will do any mode that conditions allow such as SSB, CW, RTTY, SSTV, FT8, FT4, JT65 and more which this book will cover.

You will find me on the HF bands at various times of the week working my way through as many ARRL or CQ magazines awards that I can and you may even receive a picture that I have sent via Slow Scan T.V I hope you find this book helpful and I look forward in getting you in the log one day.

73

Billy McFarland GM6DX

Introduction

The purpose of this book is to give you the reader as much information about amateur radio data modes as possible. Covering all aspects such as physical connections from the computer to transceiver, software reviews as well as how to actually have a QSO doing some form of data mode.

With so many new licensees coming into the hobby mainly due to the Covid-19 pandemic this is a prime opportunity to share some knowledge or insight into the various data modes available to the amateur radio enthusiast.

I will start by explaining various data modes and their origin. Moving onto the numerous connection methods that can be used between devices and then move onto a look at each of the different data modes. The book will then review software and conclude with QSO procedures and operating tips.

You will see that this book gets straight to the point, is easy to follow and can be referenced for returning later, when you can't remember how to set up what you have just read ;)

The Beginning

Chapter 1

The true original data mode is Continuous Wave (CW), where it is either on or off, just like where 1's and 0's are used in computer binary. However in modern amateur radio terms, CW is classed as its own mode just like SSB. Data modes seem to cover everything where a PC is used and or shift keying is used for modulation. Radioteletype (RTTY) was first used as early as the 1920's where two or more electromechanical teleprinters at separate locations were connected via radio. Information was shared between the two however this was mainly used by the military. During the First World War the expansion of these devices was great and use continued to and beyond the Second World War.

ASR33 Teletype machine

8

These devices worked by taking 1's and 0's and depending in which order they were presented would depend which letter or character would be displayed. Quite clever and originally devised by a Frenchman, Mr Baudot. Example of this type of code is below:

CODE	LETTERS
00011	A
11010	G
00100	SPACE

By using this method the combination could be almost endless and allowed the code to be set in such a way that it could take into account the possibility of errors during sending and receiving. The speed in which this information gets sent is known as the Baud rate, something which we will look at later.

In the late 1940's some Very High Frequency (VHF) amateur radio contacts were made using RTTY and at this time it was looked at by various regulators regarding band width and baud rate speeds. Transmissions were originally done with Audio Frequency Shift Keying and then later the introduction of Frequency Shift Keying, when it was approved. In the late 1950's Slow Scan TV was developed using an electrostatic monitor and a vidicon tube, introducing the ability to send pictures on air as opposed to just text as found in RTTY. As time went on AMTOR (Amateur Teleprinting Over Radio) was developed by G3PLX as a result of an International Telecommunication Union (ITU) recommendation. Which allowed RTTY to be sent more accurately having incorporated error correction and detection. Other amateur radio operators continued the development of this theory for use in Digital Signal Processing (DSP), noise suppressor systems and FAX. G3PLX then took this theory by the horns and created Phased Shift Keying (PSK) at a speed of 31 Bauds, commonly known as PSK 31.

As a result of PSK, many modes were developed from the late 1990's and early 2000's such as FSK144, JT6M, JT65, JS8, MFSK, Olivia MFSK and even the more recent FT8 and FT4. All of these have their advantages and disadvantages. Some have almost now gone extinct because of the lack of use and newer modes being created. What is apparent with development of data modes over the years is the need for an accurate, weak signal, narrow band width and fast transmission. These seem to be the common factors for modes being developed. After all the adaptation of these modes have to suit the current technological environment.

I have no doubt there are many more new modes to come in future which will rest firmly on the shoulders of the ones from the past.

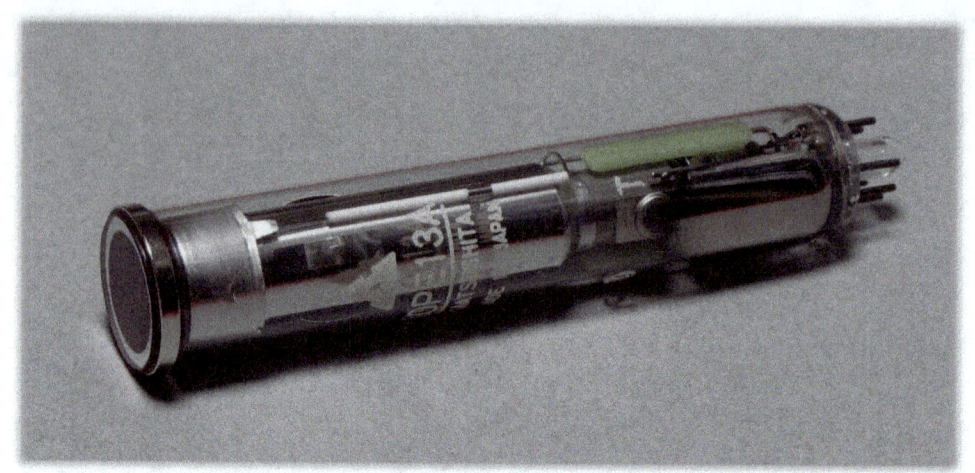

A Vidicon Tube similar to the one used in the first SSTV transmissions

A US Navy officer using a Teletype machine

The Basics

Chapter 2

As you can see from the previous chapter data modes have been developed over some considerable time. The in-depth details of those modes can be found in many places but you are very likely to be bored to death and even be put off from trying them in the first place.

I am however going to explain a few of the processes involved which should help you understand how everything works in basic terms. I will also be making reference to modern technologies so no break downs or schematics of radio teletype machines. Let's get started on some of the basics.

Frequency Shift Keying (FSK) is a modulation method which uses a pair of discrete frequency changes to the carrier signal to transmit binary code (1 is the mark frequency and 0 is the space frequency). In basic terms the transceiver is modulating the mark and space frequencies directly.

Audio Frequency Shift Keying (AFSK) is a modulation method where digital data (1's and 0's) is represented by changes in frequency (pitch) of an audio tone. The audio is usually transmitted alternatively between two tones which represent the binary number 1 (mark) and binary number 0 (space). AFSK differs from FSK as the mark and space frequencies are oscillated externally from a PC soundcard (or similar) rather than directly by the transceiver.

So is there really a difference? Which method of modulating is best? Well let's see, some advantages that AFSK offers over FSK;

- AFSK can be done on a Single Side Band (SSB) only transceiver as the audio tones modulate the carrier. The transceiver being in Upper Side Band or Lower Side Band is sufficient.
- The transceiver does not require an FSK connection in order to produce RTTY.
- As the soundcard is modulating it is also demodulating meaning it can decode or find the wanted tones from the received signal easily without you manually needing to turn the Variable frequency oscillator (VFO) to the exact correct frequency as in FSK.

Some disadvantages of using AFSK rather than FSK;

- Some transceivers do not allow you to activate the digital filters unless in an FSK mode (i.e. if AFSK is used the rig will be in USB or LSB mode).
- When used in contests and you have adjacent strong stations the received

signal can become "lost" in tuning because the software "feels" the stronger signal nearby and moves.

- You can overdrive your transceiver or cause spurious signals if you exceed the ALC (automatic level control) levels (just like when being used in SSB) but with FSK there no chance of overdrive.

Wither you use AFSK or FSK it is entirely down to your own choice and station setup. Personally I do AFSK as I have utilised this method for some years and I am comfortable with the setup. My transceiver also allows full access to filters and features in either AFSK or FSK settings.

As a result of modern technology and software development when either FSK or AFSK is being used you will be able to receive and transmit in any of the modes looked at in this book. Although slight differences are present in the setup the reality is, when choosing FSK or AFSK it is just about the practicalities of setup or install and will not hinder your enjoyment of data modes.

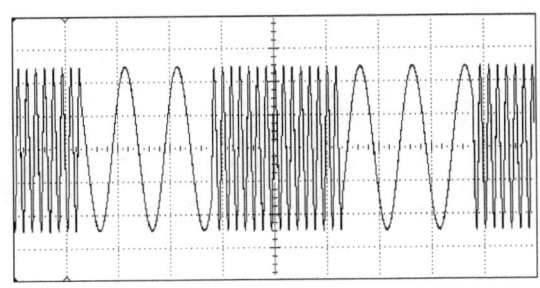

A 2 binary FSK signal

The practical set up of doing FSK can be achieved by software and physical connections however most modern transceivers offer RTTY, PSK and even FT8 usage direct by the transceiver without any other interfacing.

The transceiver can send messages and decode messages directly, displaying these received message on the screen. Some allow the connection an external keyboard giving you the option of free typing rather than using the pre-programmed messages (macros). In recent times transceivers even have the ability to log these contacts, saving them onto an SD card for upload to a local PC.

Although this method is quite neat it has a few drawbacks, mainly with user interface and mode limitation.

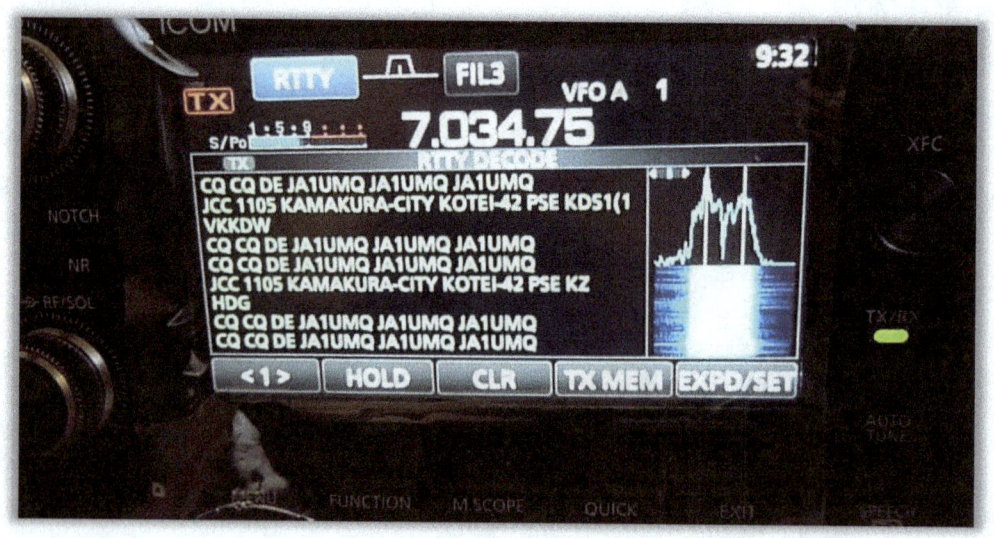

ICOM IC-7300 doing direct RTTY mode

Ideally the setup of all data modes will be done by using a personal computer (PC), some form of interface and then a transceiver. I got my three year old to draw you a sketch of the setup to help explain it.

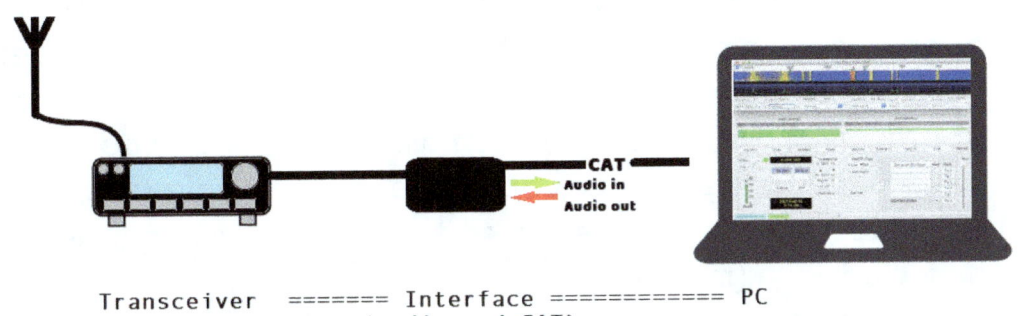

Transceiver ======= Interface ============ PC
(audio and CAT)

The two main functions of an interface is to;

1. Facilitate the audio to and from the PC to transceiver.
2. Provide CAT control, which is used for frequency and PTT control of the transceiver.

The interface can take the form of an external interface such as that provided by Rigexpert or Tigertronics. It can also take the form of an in-built interface. Most modern transceivers have a built in soundcard and CAT control which allows connection to the PC via a USB-B cable. The principle of operation for both remain the same and the only difference is physical in nature. CAT control is, *Computer Aided Transceiver,* meaning a computer aids the use of the transceiver. The software used for data modes require the CAT control and sound card to be setup separately. Some interfaces, including built in interfaces provide two COM ports. One for the CAT control such as frequency control and the other for PTT control. Some only provide one and require PTT to be conducted via CAT but more on this later. Let's compare an interface to that when operating Single Side Band (SSB). The audio (speaker connection) from the PC to the transceiver is like your brain sending signals to your mouth to speak into the microphone. The audio (microphone connection) form the transceiver to the PC is like your brain telling your ears to listen and then decode what is being heard. The PTT control from the interface is like you pressing and releasing the PTT button on the hand microphone. It really is as simple as that. To summarise in order to do any form of data mode you require the ability to convey audio between the PC and Transceiver and you need the ability of frequency control and to put the rig into transmit via the PC.

Rigexpert Ti-8 interface

The Connections

Chapter 3

In this chapter we are going to look at the physical connections between various transceivers and your PC. Before I get started I want to clarify that most modern computers have a USB port, as such cable designs will be based on USB connectivity. The original COM ports can still be found on PC's but are far less common.

Starting with **CAT control** if your transceiver is lucky enough to have a built in USB – B port for CAT control then all you need is a USB –B to USB cable (also known as a printer cable).

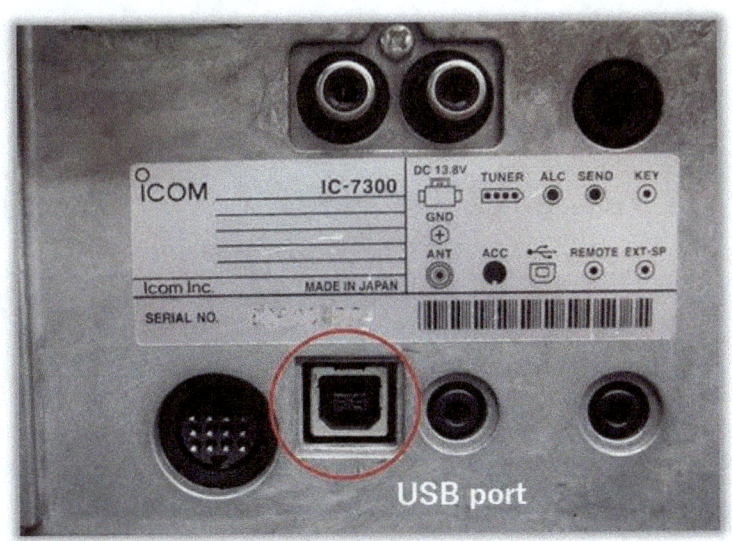

ICOM 7300 rear panel

Most modern transceivers Such as ICOM IC-7300, IC-7610, Yaesu FTDX10, FT-710 AESS, FTDX101, FT-891, Kenwood TS-590. TS-990 and so on, will have them. Now you need to be careful as most of these transceivers also offer a soundcard (audio in and out) via this connection *BUT* not all of them so you will need to check with your manual.

> 🖱 Mice and other pointing devices
> 🖵 Monitors
> 🖧 Network adapters
∨ 🖳 Ports (COM & LPT)
    🖧 Communications Port (COM1)
    🖧 Communications Port (COM2)
    🖧 Intel(R) Active Management Technology - SOL (COM3)
    🖧 Silicon Labs Dual CP2105 USB to UART Bridge: Enhanced COM Port (COM6)
    🖧 Silicon Labs Dual CP2105 USB to UART Bridge: Standard COM Port (COM8)
> 🖨 Print queues
> 🖨 Printers
    🖨 P

What the COM port looks like in device manager on a windows PC.

The above image shows how this type of connection appears to the computer (located within device manager). Like all of these connections that will be discussed in this book it will show as a COM port. This example is a Yaesu FTDX10 which has a COM port for CAT (frequency control) and a COM port for PTT control (6 & 8).

Ok, so what if the transceiver you have does not have a built in COM port chip etc. Well you will need to make or purchase some form of interface.

A simple interface to make for all Yaesu transceivers which have a 8 pin mini din socket (usually tuner socket or similar) is seen below;

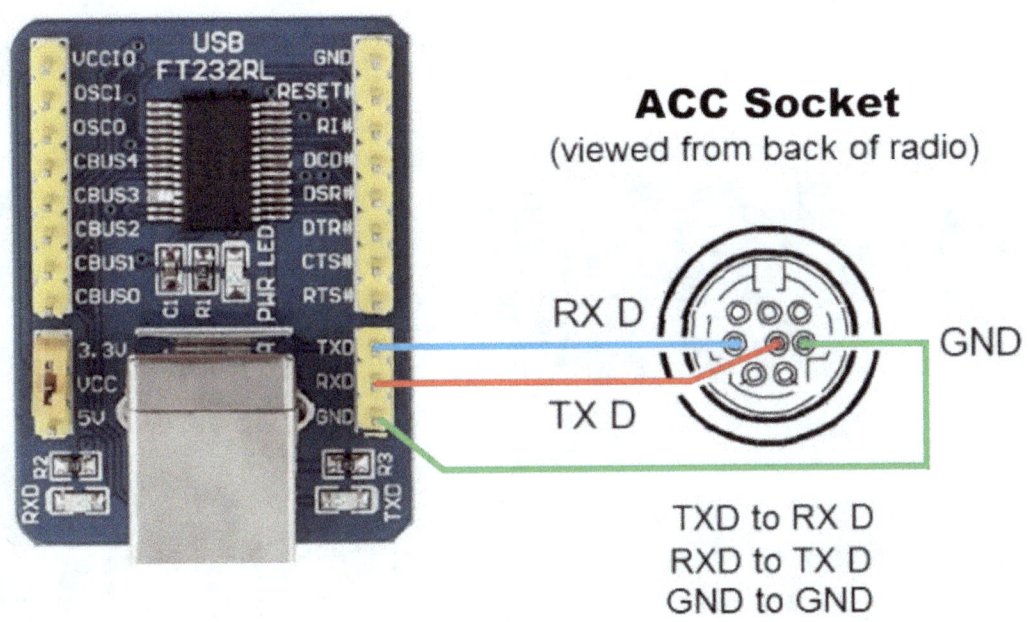

USB FT232RL UART with 3 wires going to an 8 pin mini din socket

The above image shows some form of universal asynchronous receiver transmitter (UART) converter. These allow physical wires to be connected at one end and a USB-B cable at the other. It has a built-in chip which provides the connectivity to the PC. The one pictured costs around £5 at time of writing and has a FT232RL chip but CP2105 are also common.

The transmit data (TXD) from the PC goes to the received data (RXD) connection on the transceiver and vice-versa. The ground (GND) connections are connected to each other. This example only shows an 8 pin mini din however the socket on the Yaesu transceiver could be a 10 pin din, db9, db25 and 8 pin din (not mini). What you need to look out for are the sockets which have TXD, RXD and GND connections.

UART with a 1N4148 diode and 3.5mm stereo jack

When it comes to ICOM it is slightly different. The diagram above shows you how to wire an UART board to 3.5mm stereo jack plug.

Both the TXD and RXD get connected to the tip of the 3.5mm stereo jack (red and blue wires), the grounds are connected straight through to each other (green). As both TXD and RXD are connected to the same point (tip of 3.5mm) then a 1N4148 diode gets connected in line with the TXD connection.

The 3.5mm stereo jack gets plugged into the ICOM CI-V socket on the rear of the transceiver and this will afford you CAT (and PTT) control

Alternative CI-V cable using a 10K ohm resistor

Like everything in life, you will find more than one way to do things. The above image shows an alternative method of making an ICOM CI-V cable. Where a 10k resistor is used at the VCC connection and again also connects to the tip of the 3.5mm stereo jack plug. I have had success using both methods.

When it comes to other transceivers it can get a bit funny. Some transceivers like straight through connections for example RXD – RXD and TXD – TXD. Others like a cross over such as TXD – RXD and RXD – TXD. As well as this others also require the clear to send (CTS) and request to send (RTS) to be connected together. At this point homebrew interfaces will require a bit of "trial and error" to get the connections' working the way you want.

Kenwood transceivers are also not exempt from this varying between models. I have tried a

few different variations over time but
generally speaking the next diagram should
work for you.

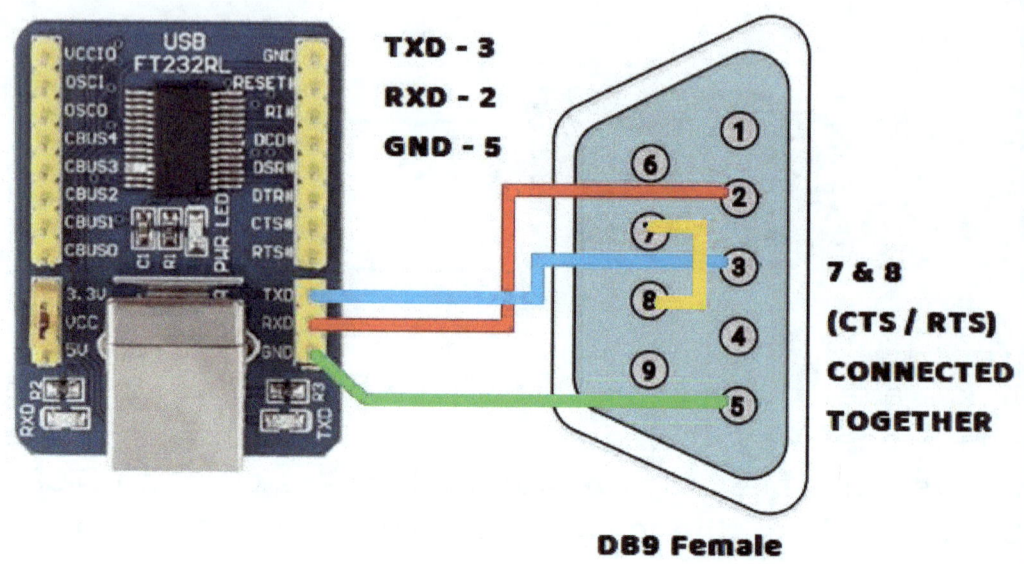

TXD - 3
RXD - 2
GND - 5

7 & 8
(CTS / RTS)
CONNECTED
TOGETHER

DB9 Female

UART to a DB9 female connector. Note RTS/CTS connection.

In general terms these interfaces are not
difficult to make. You only need some cable
and even if you are not great at soldering you
will be able to solder onto the UART no problem
as they are robust and have plenty of room to
get the soldering iron tip in-between pins.
Don't forget to include some hot glue to help
the wires from being pulled or displaced when
fitting. I have provided these diagrams for
reference and illustration purposes.

**WARNNG: Any homebrew construction always comes
with some risk that you may damage your
transceiver.**

Now that is the CAT and PTT (as you can PTT via CAT) interfaces covered we will now look at the basic homebrew soundcard interface.

Most PC's will have a built in soundcard. It will either be a speaker and microphone jack which is 3.5mm or a combined 4 pole 3.5mm headphone jack. Although connections might differ the principles are the same, speakers for audio out and microphone for audio in.

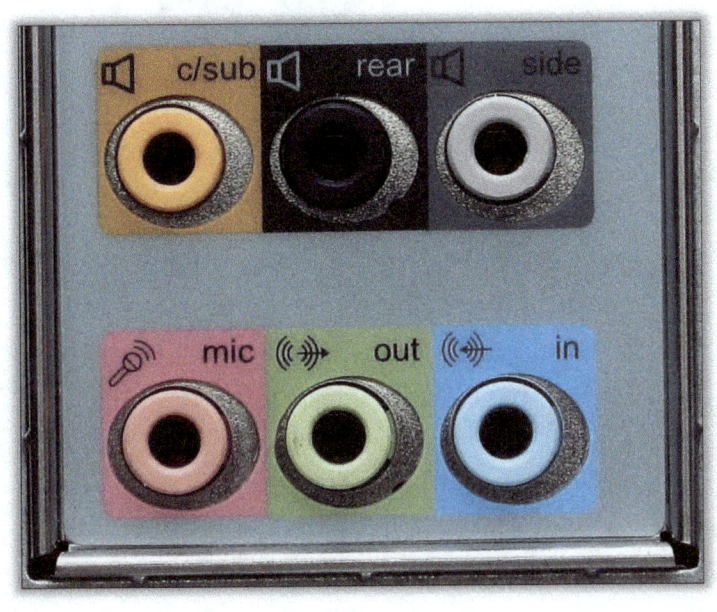

Modern PC has more than one option – Mic/ Line IN and Speaker / Line out.

We are not going to make a soundcard so should your PC not have one then you will need to purchase one. You can use a cheap USB soundcard (plug into a USB socket) as pictured, they do work and performance has certainly improved over the years.

Typical USB sound card available for under £2

You can see from the picture that the USB soundcard just provides a 3.5mm stereo socket for microphone and speaker connections. So it should come as no surprise that you will need to have some sort of connection that goes from a 3.5mm stereo plug to your transceiver (unless you want to open the plastic enclosure and solder directly onto the PCB).

You can just use a direct connection between both but there is a chance that some interference generated in the PC can travel back towards the rig (and vice-versa). To help prevent this you want to ensure that the audio lines are isolated from each other. One method of isolation is by using a transformer.

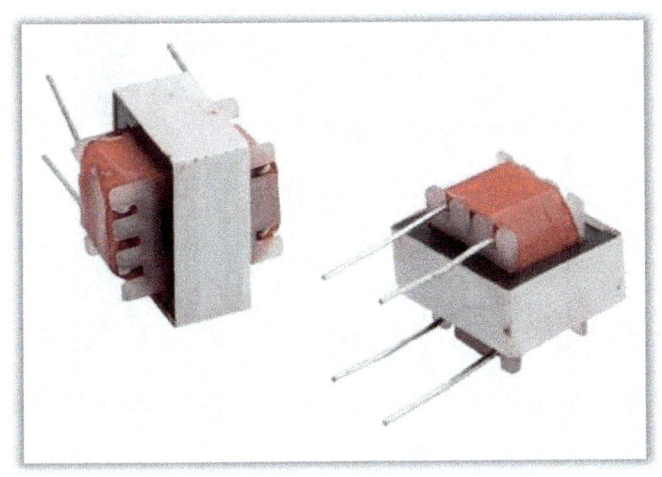

600:600 OHM Audio Isolation Transformer

The audio line and ground gets connected to one side of pins. The continuation of these wires are connected at the opposite side of pins. By using these transformers it is ensuring that the audio lines are isolated and will provide a clean audio signal for modulation during data mode usage. The direct audio levels from the PC speaker to the transceiver may be too much for the transceiver to handle. Although you can control the speaker volume on the PC the levels could still be too strong.

This could cause problems later when using software to modulate the signals. In order to help combat this we will need to use a variable resistor in line with the audio from the transceiver to the speaker connection.

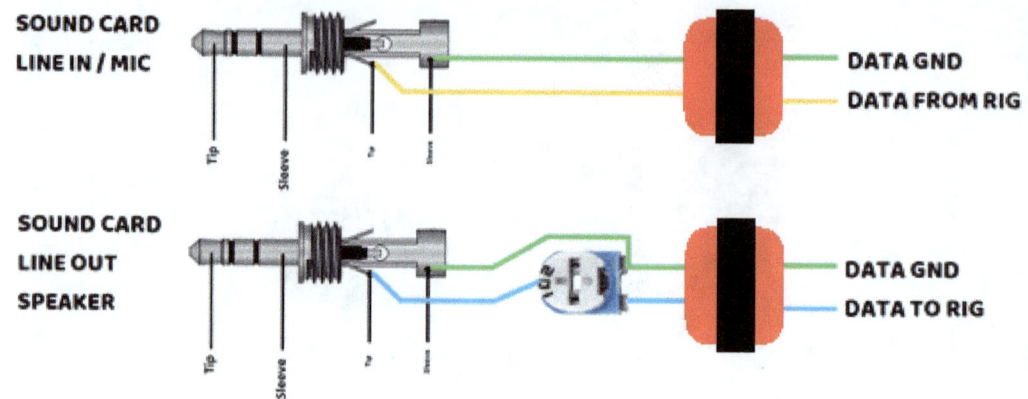

The above shows the practical connection of a pre-set resistor and audio transformers. This form of interfacing has been around for some time and you can even purchase easy to assemble kits that just require some soldering. These kits can also offer DTR, RTS (for computer keying) as well as the audio connections.

An Easy Digi interface for data modes.

This interface does not have a pre-set resistor but is constructed to reduce audio levels as well as protection for the connected devices.

At the transceiver connection, it will vary between models and manufacturers. For Yaesu it is usually a 6 pin mini din connector labelled DATA.

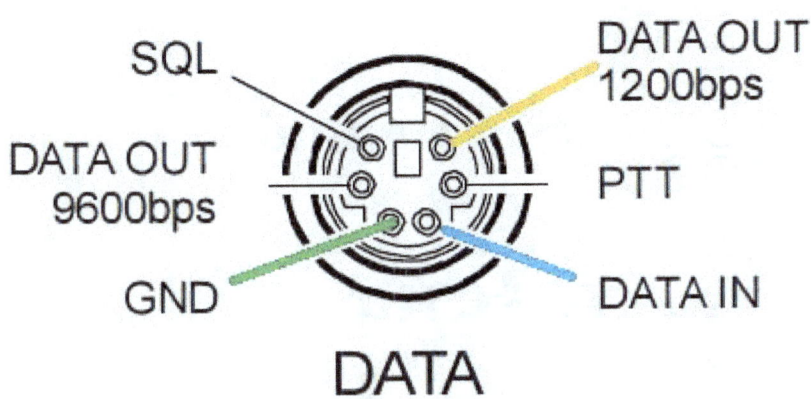

Using the same colour code as previous images. Green is GND, blue is audio from the PC speaker to transceiver and orange is audio from transceiver to PC microphone.

You can purchase a 6 pin mini din cable with moulded plug already attached. That way you are only cutting the wire for connection to the transformers as soldering the mini din connectors can be tricky. You need to remember that by using rear data sockets the chances are that your transceiver will need to be in USB-D/ LSB-D or DATA mode to allow the radio to see the PC audio at this socket for modulation. This will likely be an FSK transmission.

If the radio is in USB/LSB only the chances are it will be looking for audio via the front microphone connector. You can of course connect to this socket and this will be AFSK transmission.

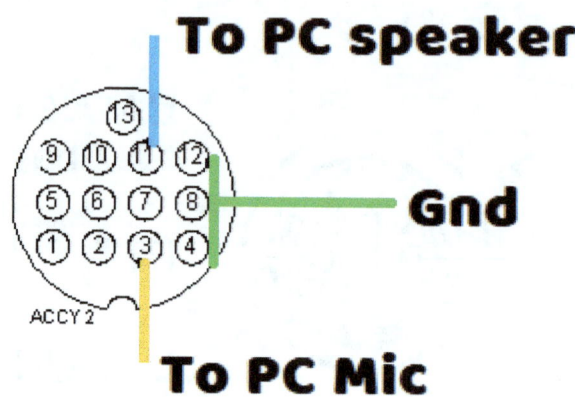

View of the 13 pin din Kenwood connection

On most Kenwood radios you will find a 13 pin din connector. This connector allows access to receive and transmit data signals at the transceiver. Common wiring of these pins can be seen in the image above. For most ICOM models connection can be made via the 8 pin din socket at the rear of the rigs.

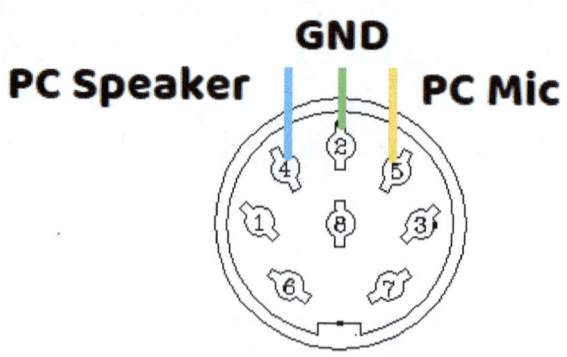

ICOM 8 pin din plug data wiring

You need to be mindful if using rear data sockets that the transceiver will need to know that it has to use these sockets. This is usually done via an internal menu.

You can modulate the transceiver via the front mic connector. This allows AFSK transmissions. You will however need to set your rig in USB/LSB mode. Also it is similar to using a hand mic you must ensure that the output volume of the PC doesn't exceed the ALC levels on the transceivers meter. If it does, just turn down the speaker volume on the soundcard that you are using.

An example of wiring for ICOM 8 pin mic socket is below.

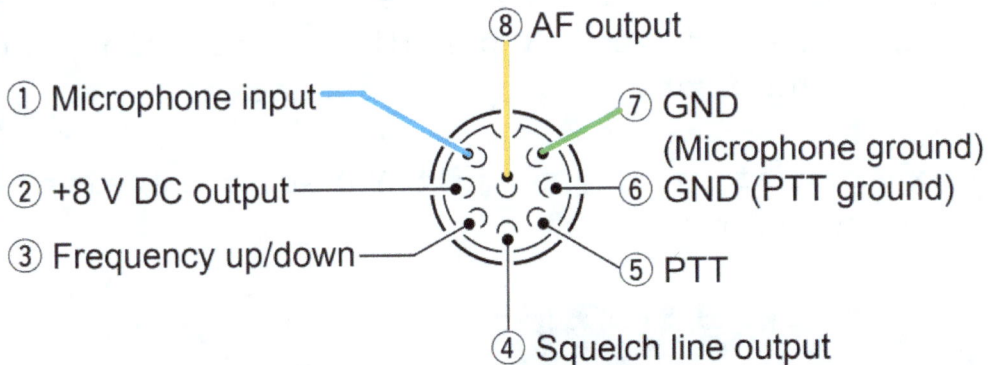

Front panel view

Mic wiring for ICOM showing same colour reference for connections. Note how AF out is provided.

Note how ICOM offer an AF (audio) out option via the front mic. Most other manufacturers do not and if you are using the 8 pin mic connector for audio in then you may need to use the rigs external speaker connection for audio out. That covers most of the audio connections available or needed.

I have one more thing that I want to cover and that is the construction of a homebrew PTT device. Let's say you cannot PTT (press to talk / usually referred to when rig has to go into transmit) your transceiver via CAT control. Then you will need some form of ability to PTT the rig via computer. To do this we can just use an UART converter which has the built-in chip for PC connectivity. We then need to connect this to a socket which allows PTT. Depending on your transceiver this may be via a rear Phono socket, 8 pin din, 6 pin mini din or even the front 8 pin mic connector.

An example of wiring to suit Yaesu transceivers is found below;

UART Converter with a diode between GND and DTR. The DTR or RTS can be used for PC keying. In this example DTR is connected to a Yaesu 6 pin din DATA socket.

Unlike CAT we are not using TXD or RXD we are using Data Terminal Ready (DTR) or RTS for keying. Most if not all software for data modes have the option for you to select DTR or RTS for PC keying. Common practice is to use RTS for PTT and DTR for CW. The principle is the same no matter what option you use.

To summarise this chapter; you will have learned that you need an audio interface which allows audio isolation from the PC to transceiver. You need an interface which will provide CAT control. You may also need an interface which performs PTT control of the transceiver via the PC. All of which has been covered in this chapter.

From reading this chapter you should have gathered that using a modern transceiver which has one single USB-B connection that provides PTT, CAT and Soundcard connection via this USB cable is the best possible option.

Far less construction is needed and setup is likely to go more smoothly compared with 2 or 3 separate homebrew interfaces.

Remember homebrew construction comes with risks. If you are not prepared to take the risks then do not homebrew your own interfaces or other devices!

The Pro Products

Chapter 4

Professional built products are available for those who do not want to construct their own interface. Several are on the market but the principle of function and operation are the same for all makes.

The external interface will provide the ability for CW keying, FSK and AFSK keying, sound card for audio as well as CAT control all in the one unit. Some of these interfaces even have Wi-Fi that allow remote station operation. Generally there are some form of interface cable required that allows connection to your transceiver via various ports.

A Rigexpert interface

The one pictured is an interface by Rigexpert who also make excellent antenna analysers. This interface has a built-in sound card with volume mixing via the rotatory knobs on the front. Audio in (1 & 2) and once mixed the output interface volume can be controlled by the out rotary knob. These volume controls on the front make it very easy for adjusting the levels of audio on the screen and output without needing to adjust the volume on the actual PC. This is very useful for when doing numerous modes or using various software during the one operating session.

You can purchase an interface cable for your transceiver model however Rigexpert publish wiring diagrams for all their interfaces showing you how to wire the connectors relevant to you. If you have the ability then it is always easier and most cost effective to make your own interface leads.

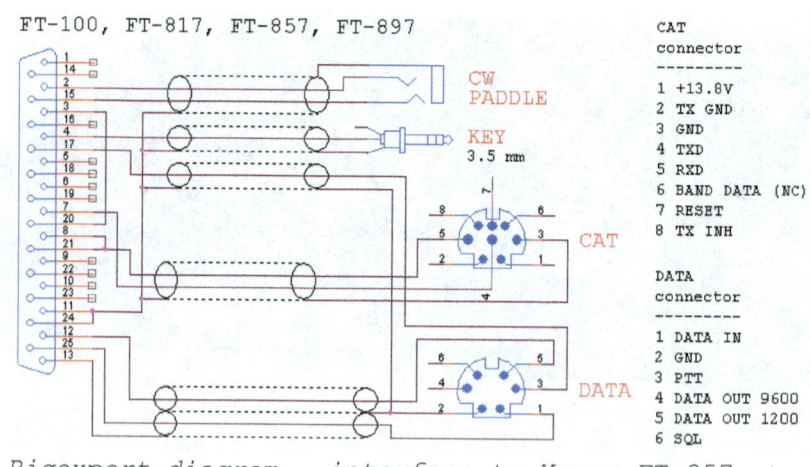

Rigexpert diagram - interface to Yaesu FT-857 etc.

As well as the classic big names in amateur radio interfaces that you see advertised everywhere you also have individual amateurs who make products at a professional quality and to a high standard.

The traveller interface by G3LIV

The one pictured is built by G3LIV (or produced by G3LIV designs). It is a professional built interface with the same functions as the Rigexpert interface. The main difference with this interface is that it is hard wired for use with the one transceiver type. Meaning should you purchase a new rig you would likely need a new interface to match it.

Full product support is available by means of a manual for download as well as driver software for connecting it to your PC. Overall these type of amateur pro products offer a lot at about 1/3rd of the price. If you are interested in only receiving (as a short wave listener) data modes these interfaces will still work for you.

You can download applications for your mobile phone. The microphone on the mobile phone will pick up the audio tones from your transceiver's external speaker and decode the message displaying this on the screen.

This can be a cool way of getting your kids or other youngsters into the hobby especially when the international space station regularly sending out SSTV pictures which can be picked up on a simple 2M / 70cm hand held transceiver.

In short if your transceiver does not have a built-in CAT control and soundcard and you are not very good at construction then you will need to purchase a professional produced interface should you want to do any type of data mode.

Mobile phone decoding using an application

The Modes

Chapter 5

We are going to look at each data mode and how they function on a brief basis.

PSK – Phase Shift keying is where we send data by means of changing the phase of a constant frequency. Modulation is done by varying the sine and cosine inputs at a particular time. PSK is found in Bluetooth and RFID products.

BPSK – Binary Phase Shift Keying is the simplest form of PSK as it uses two phases and 180 degrees separation. This means it can deal with the highest of noise levels as the demodulator only has two constellation points (180 degrees apart) to deal with, before the incorrect decision is made. This makes it the most robust of PSK options.

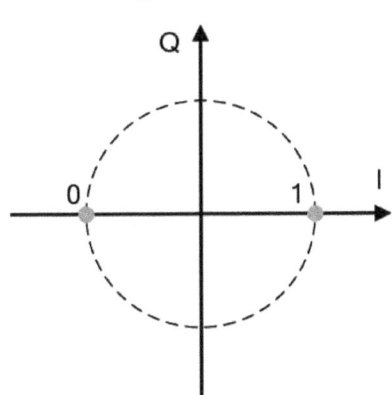

Example of constellation diagram for BPSK

QPSK - Quadriphase Phase Shift Keying is similar to BPSK however it uses four phases rather than two. This means that it can be used to double the data rate (over BPSK) but still occupy the same bandwidth.

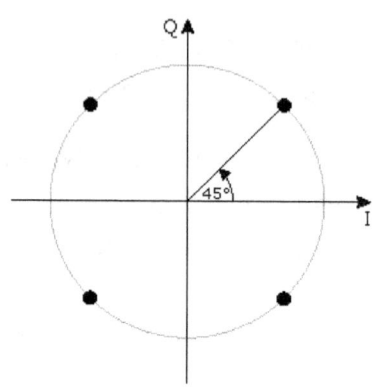

Example of constellation diagram for QPSK

Software such as Mix W allows you to select either BPSK or QPSK as a modulation type. Generally I use BPSK however with the information above the choice is yours.

PSK 31, PSK 63 and PSK 125 all function on the same principle the difference is the bandwidth of transmission. PSK 31 is the narrowest of transmissions and is also the slowest. Remember the wider the bandwidth the more data that can be sent so therefore the faster the transmission. You can see me doing these modes on my *Youtube* channel, for visual learning.

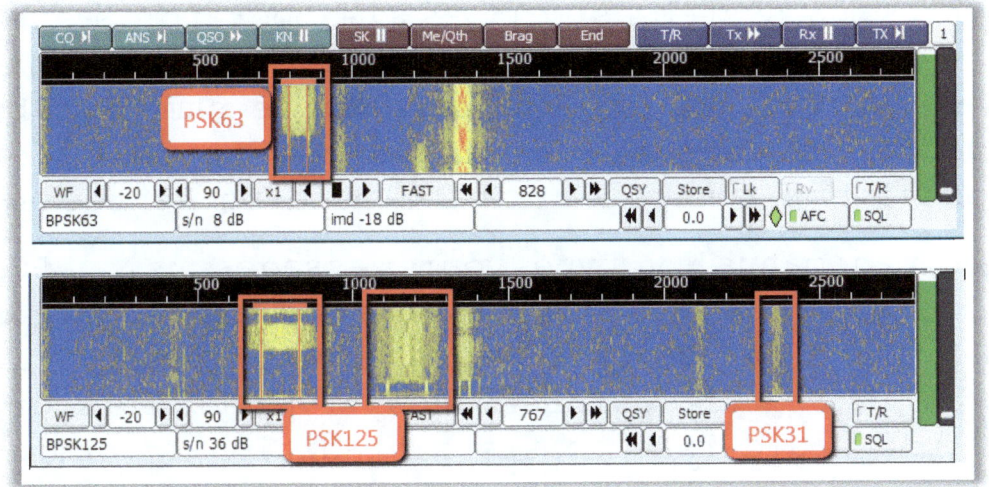

On air you will find PSK 31 and PSK 63 regular at the narrow data portions of the bands. If you are unsure of where these are on each band then check this website out;

https://www.qsl.net/sv1grb/psk31.htm

PSK transmissions are not fully automated like FT8 so require your engagement for *every* QSO.

Find a spot on the relevant band and select a CQ macro and give out a call. Be sure to use capital letters for CQ, Q codes and call signs. Once you give a few calls you will hopefully see a reply which will look like a single line a few millimetres wide (see images above).

Click on the line to decode it (well most software works that way) and enter the call sign of the station into the software log. Process your way through the macros. In PSK 31 you generally find allsorts being transmitted from transceiver type, antenna, software and operating system as well as the operator's age. The mode is slow and it really is just like rag chewing but on data modes. The same can be said to a degree for PSK 63 but it is a lot faster and sometimes can be used in contests.

You can manually type your messages using PSK 31/63 but unlikely to be able to type for PSK 125. This is one major advantage of using this mode is that it offers manual typing. In short if you would like to try a traditional form of data mode that is not automated and requires human participation then start off with PSK 31.

MFSK – Multiple Frequency Shift Keying is a slight variation of Frequency Shift Keying which uses more than two frequencies. It is a form of digital modulation but rather than transmitting one bit of data it transmits two or more simultaneously. Some of the most common MFSK types are, *MFSK 8*, *MFSK 16* (developed by ZL1BPU) and *Olivia MFSK*. It works on the same principle as PSK 31 and transmission type and style are also similar.

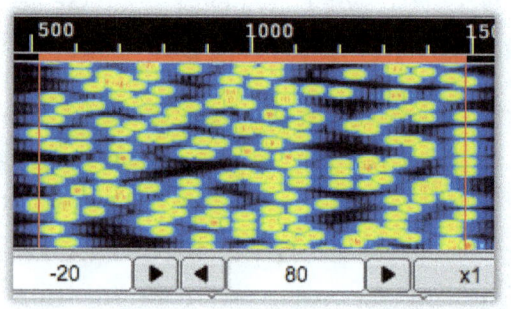

MFSK Oliva signal on software

41

RTTY – Radioteletype is a form of Frequency Shift Keying with the speed (baud) rate usually being 45.45. However 50,75,100,150 and 300 are known. The frequency carrier shift for amateur transmissions are usually 170Hz and these settings can be checked in the software that you are using.

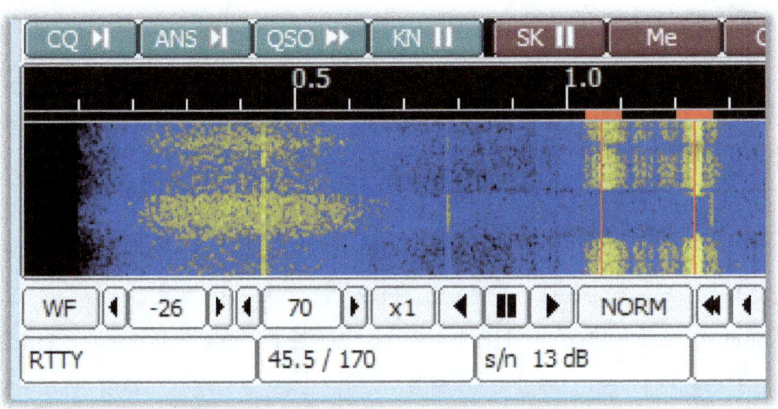

RTTY signal showing on software

RTTY has a unique audible tone which once you hear it a few times you will instantly recognise it on air. Depending on which software you are using will depend on how the QSO is engaged.

Most software allows you to click on the RTTY signal which looks on screen like two lines (image above) with a space between them. The space distance will vary depending on the Hz used as previously stated but most amateurs use 170Hz.

You can manually put the software into transmit and type your message. If you are slow at typing the audio sounds like a stutter and the two audio tones on screen are broken, meaning that it is not modulating because there no characters to send. If you are not good at typing then use macros. This mode was originally designed with machines that people typed into so I find manual typing enjoyable. Working on the same principle as PSK it requires human interaction in order to process through the QSO. RTTY is used in either AFSK or FSK and found regular in contests. The *CQ WW WPX RTTY* contest is one of the most enjoyable contests you can do. You can participate with friends and still chat about every day stuff. Have a beer at the same time and there is still a skill in working stations.

RTTY with a shift of 170Hz and baud rate of 45.45(standard amateur settings) requires about 250Hz of bandwidth. So a 400Hz or 500Hz wide filter (on the transceiver) is better to be used during contests. This is because you will see that the quantity of stations are so great that nearby strong stations cause QRM, bad decodes and on occasions can block you from receiving anything. You will even see RTTY stations transmitting over your signal, so basically in between your shift.

When it comes to filters there a few things you need to understand. What I have found over the years is that there is a relationship with filter width of the rig, volume input levels to the PC and decode ability. If you set your PC volume levels with the rigs filter at say 3 kHz and then you change your rig to 500Hz you will notice that the software becomes overloaded with signals or noise and it will need adjusted. Likewise if you set your PC volume levels with the rig at 500Hz wide and change the rig to 3 kHz you will notice that hardly any signals are being decoded or even making onto the screen. When switching between both you can see the problem specially if you are switching regular (now you can see how useful those external interfaces are with volume control knobs on the outside).

So I personally find that when I am scanning the band's I will set up the PC levels to suit a 3kHz wide rig filter and during CQ calls (run station) at contests or for general contesting I will set up the PC levels to suit a 500Hz wide filter. The same principle can be applied to most data modes but generally RTTY is the main data mode used in contests.

When trying to run a pile up or work that DX station you will find that even in RTTY mode they operate split frequency. That is they will transmit on one frequency and listen on another. The general rule of thumb is that they will listen for people calling at least 1 kHz up. This can be quite tricky to set up on some software even although the rig is set to transmit on the split frequency. You will also notice that the frequency displayed on the rig differs from what is on the software (screen). That is because the software shows the actual transmitted frequency (after the shift) so any spotting or reference to actual frequency should be taken from the software and not directly from the rigs VFO. Overall RTTY is by far in my opinion the better data mode to do. It requires a skill to have a quick QSO rate especially during contests and is enjoyable to do, due to the level of human interaction needed.

SSTV – Slow Scan Television was created in the 1950's where we can send an image over the airwaves. It uses a form of FSK and can have a bandwidth of up to 3 kHz. There are lots of modes but the most common and image sizes are detailed below;

Family	Developer	Name	Colour	Time	Lines
Martin	Martin Emmerson - G3OQD	M1	RGB	114 s	240[1]
		M2	RGB	58 s	240[1]
Scottie	Eddie Murphy - GM3SBC	S1	RGB	110 s	240[1]
		S2	RGB	71 s	240[1]

Note the Scottie mode was developed by a Scottish operator hence the name *Scottie*. Each mode varies in pixel, image size and length of transmission. SSTV is found in the voice portions of the band due to the large bandwidth. Unlike most data modes where you transmit over a range of frequencies SSTV is found at specific calling frequencies, like channels. The details of these can be found online. Although most transmissions are in upper side band you can also find SSTV signals on lower side band and FM. The international space station regularly transmit SSTV on the 2m/70cm band. The QSO process is similar to most modes you send a picture with CQ text on it, await a reply then send another image with exchange details on it.

The cycle of process continues until the end of the QSO. SSTV is a good mode for getting started in data modes and is not time pressured so you will have plenty of time to get used to the software and QSO process.

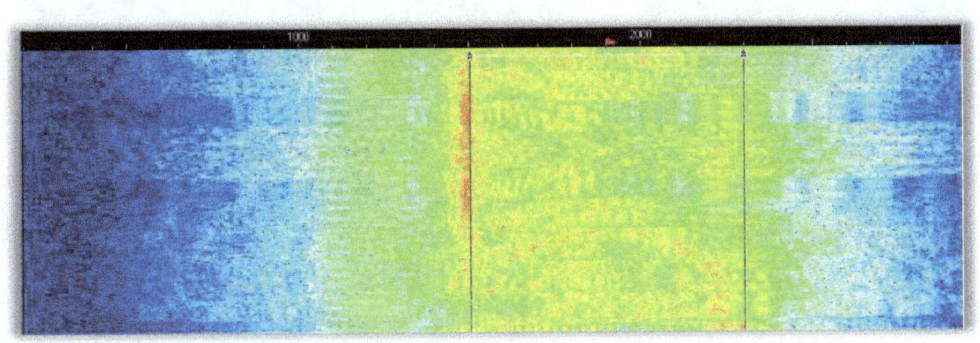

SSTV signal as it appears on MixW software

FT8 – Franke Taylor 8FSK (named after the developers Joe Taylor and Steve Franke) is a weak signal form of FSK and is around 50Hz wide. It uses the communication protocol that is an automated, bare minimum of information needed to complete a recognised QSO. So therefore it is not real time chat (keyboard to keyboard like what RTTY and PSK can be.

FT4 – Franke Taylor 4FSK works on the same principle as FT8 however the transmission time is reduced by roughly two and half times but at a cost of being about double the bandwidth as well as being less sensitive to weak signals. However it is about 10dB more sensitive than what is needed for an RTTY contact to be made (even more reason why RTTY is the more skilled mode to do ;)

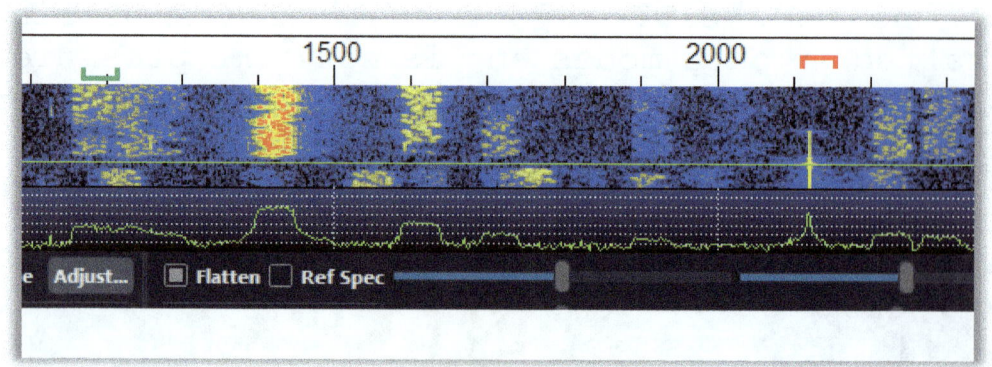

FT8 signals at various strengths

Both FT8 and FT4 (and a few more such as JT9) require the use of synchronised timed transmissions. Working from the computers clock a transmission is made for a timed period. So you will transmit for a period of time using the pre-set macros. Then you receive the other station for a period of time before you then transmit again. This cycle continues until each of the messages can be sent to make the QSO. The overall time of transmission and receive period will vary depending on what mode is being used. You will transmit on either every odd or every even timed period.

This is one major advantage that FT8 has over these traditional timed modes such as JT65 is that it operates on the same if not better sensitivity levels but far quicker exchanges.

JT65 – Joe Taylor 65 tones works on the same process and principles as FT8 and FT4 but transmits its cycle on the UTC minute (either odd or even transmit cycle) and lasts for a duration of approximately 48 seconds. This was the original weak signal mode but with the creation of FT8 and FT4 the demand for JT65 is not there.

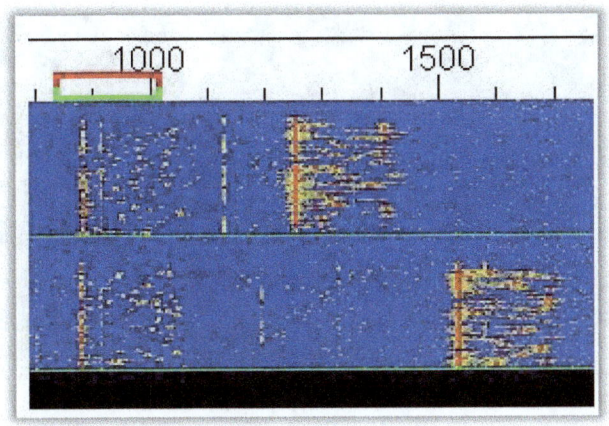

JT65 Signal on WSJT-X software

JS8 – Jordan Sherer created the JS8 call which was originally named FT8 call using the FT8 principles. It turns the automated messages as used in FT8 into the ability to have a QSO by keyboard to keyboard typing. You can send your message to another operator or group of operators for them to check there and then or at a later time simply replying when ready. Like a form of text messaging or emails but using HF propagation. Bandwidth can vary from 25Hz to 160Hz dependant on transmission speed.

It has lots of pre-typed messages for selection as well as a status notification to let people know you are available for live QSO's. This is a process which is a mix of PSK 31 and FT8 and people generally use this for the keyboard interface and the ability of weak signal decoding.

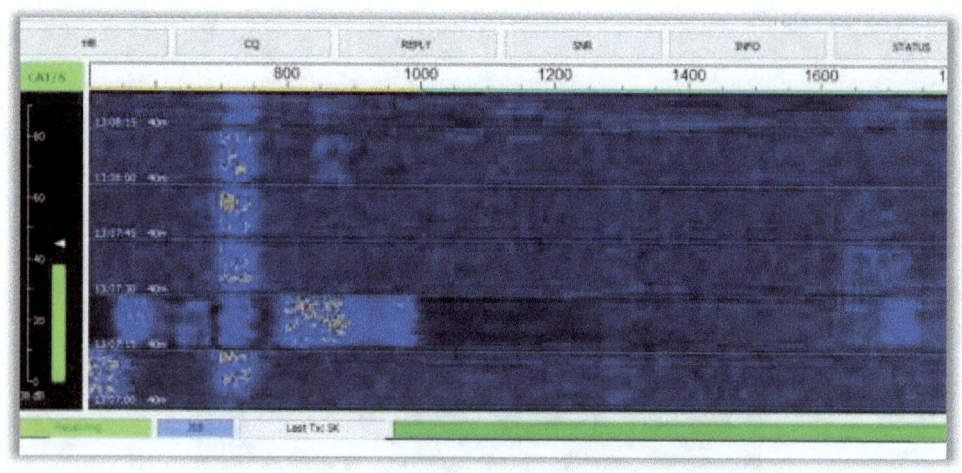

JS8 signal in JS8CALL software

CW – Continuous Wave is one of the earliest forms of data modes where the signal is either on or off. It is made up of dots and dashes, where a dash is the same length of time as three dots. You can send these by a hand Morse key or with a Morse paddle where the dots are sent on one side of the paddle and the dashes on the other. When using a PC to send and receive Morse there are two main methods of transmitting. You can send using CAT protocol with DTR keying via the CAT connection. Or you can make up an interface where it uses DTR keying but connected to a stereo jack plug, to fit into the Morse key socket on the transceiver. Note I said stereo and not mono. There has been issues in the past when a mono jack is used. A solid tone is sent only when trying to key from a PC so use a stereo jack. As well as a stereo jack please ensure your transceiver is set to a straight Key and not as a paddle.

As the straight Key setting is selected on the rig then the CW sending speed will be controlled by the software but the CW tone will be controlled by the rig. However if CAT is used for CW keying then on occasions the software controls the CW pitch or tone pitch of Morse code.

Dependant of software and connections some filters can be controlled by the software but failing that narrow filters can be set on the transceiver. 500Hz or 400Hz wide filter should be sufficient and make sure the rig is in CW or CW-R (upper side band or lower side band) usually it is in CW.

As with all other data modes you need to adjust the input or microphone volume on the software to ensure you get a nice signal on the screen for decode. You don't need to worry about output volume for CW and your ALC will always be burst.

A CW signal showing on MRP40 software

CW using software allows direct typing into the keyboard to send or by using macros, as with PSK and RTTY. There a few people who will say "you're not a real CW operator because you are not decoding with your ears". Which is true to a degree however this is a great way of getting started into the mode or if you are only making a few contacts a month.

If you are looking to develop a skill set in data modes then I recommend you starting with RTTY.

If you are content with sending limited messages with little human interaction or have a small powered station then I recommend you start with some FT8.

As with all data modes the ones covered here are full duty cycle and you should check your transceivers specification prior to use. You could quite easily burn out the transceivers power amplifiers with incorrect usage.

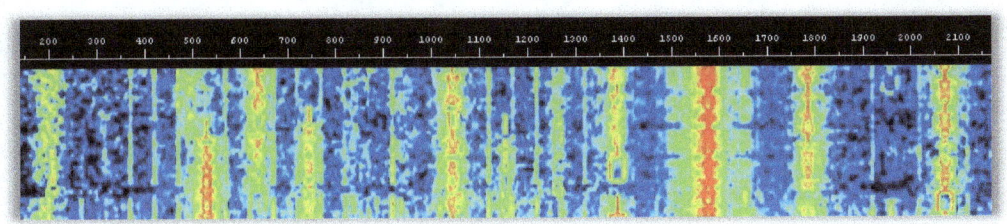

Various data modes being decoded on software

Once you understand the basic idea of operation and the basic understanding of how any data mode works you will be able to setup and try all modes very quickly.

The only thing that really changes in the development of data modes for the user is the software interface. The "behind the scenes" of data modes is never for the average amateur. All that amateur radio operators need to worry about is how to set up the software, the physical connections of the equipment and the process of the QSO.

The Macros & QSO

Chapter 6

Traditionally data modes permitted the custom sending of messages by the operator simply typing these in manually. Today you can still do this in certain modes such as PSK or RTTY. However the modern world relies on some form of automation and the most common way of doing this is by using Macros.

Macroinstruction (macro) is basically a programmable pattern that makes use of a particular imputed sequence to produce a pre-set output sequence. In other words it makes tasks less repetitive by the simple click of a button.

Although each and every software will have a different way in which you set up macros the principle will be the same. Let's start with a typical CQ call macro.

<TX> CQ CQ CQ de GM6DX GM6DX CQ pse K <RX>

The first thing we need to understand about macros is that we are also telling the computer what to do with the transceiver. In the above example <TX> or {TX} is letting the computer know we want the rig to go into TX. Then we send the text and finish with <RX> or {RX} where we are telling the computer to put the rig back into receive. Generally speaking commands are in some form of brackets. Not all software requires the inclusion of TX and RX however the brackets of other tasks such as CALL or RST are generally needed by all software.

The start of the transmission can be missed on occasions. This is due to the delay in people physically clicking on the signal they want to meaning they miss the start of transmission. In order to combat this you will see some operators starting transmissions with a few dots for example, ...CQ CQ CQ de GM6DX. This way there no issues should you miss out decoding of a few dots. After CQ we send *de* (meaning from) our call sign and then finish up on *pse* (please) *K* (reply).

Contest macros are different in respect that the text may relate to a particular contest an example;

...CQ Test de GM6DX GM6DX test

The word test being short for contest. A lot of the abbreviations are found in Morse code and generally you will be able to learn as you go rather than a particular task before transmissions of any data mode. Once we put out a CQ call we will wait on a station sending their call using a macro similar to

{TX} OY2T OY2T {RX}

We will then decode their call on our screen (hopefully) and now need to create an initial reply.

{TX} {CALL} de GM6DX

TNX FER UR CALL

RSQ: 599

LOC: IO75XV

NAME: BILLY

QTH: AIRDRIE

So how cpy? {CALL} de GM6DX KN {RX}

Looking at the above message we have told the rig to go into transmit. Then we send a call sign {Call} this is the call sign that you have highlighted or that is sitting in your log (the station who is replying to your CQ call). The remaining is general exchange text, RSQ (RST) Locator, operator name and QTH. To finish this exchange we again send the other stations call sign ending with the rig going into receive.

If this was a contest, an exchange could be

{TX} {CALL} 599 {NRS}{NRS}{CALL} {RX}

Transmit, stations call sign {CALL}, signal report, contest exchange x2 {NRS}, stations call then receive. The reason we send the station's call sign twice (once at the start and once at the end) is that they might miss the one at the start which is very common with

QRM during contests, so they will decode it at the end.

Now we have received the other stations reply we need to send out a final message something similar to;

<TX> <CALL> de GM6DX

TNX OM FER NICE RPT

Full details of my call found on QRZ.com

Best 73 from Scotland

<CALL> de GM6DX SK N <RX>

Again nothing new here in relation to commands.

A contest final could be;

<TX> <CALL> TNX GL GM6DX TEST <RX>

What text you put in a macro is entirely up to you. Some people type a set of macros for doing CW, RTTY and PSK. As the content will vary. The slower modes such as PSK 31 generally have a lot of text and full words that if you sent in CW the other station would leave half way through the QSO. It is also said that by using lower case letters the message is quicker at being sent compared to capital letters. I am not completely convinced you notice any real difference however please put CQ, all Q codes and call signs in capital letters when typing or using macros.

Some common macro commands are listed in the table below.

MACRO	MEANING	MACRO	MEANING
<CALL>	Call of station	<NRR>	Exchanged received
<MYCALL>	Your call sign	<NRS>	Exchanged sent
<QTH>	QTH of other station	<TIME>	Insert current UTC time
<RSTS>	RST to send	<QSOBEFOREDATE>	QSO before on this date
<RSTR>	RST received	<MODE>	Send current mode
<QSONR>	This QSO number	<MILES>	Distance in miles

Some macro text that you could use for each mode:

CQ all modes;

CQ CQ CQ de GM6DX GM6DX GM6DX CQ pse K

CW Reply;

<TX> <CALL> de GM6DX TNX OM UR RPT = 599 NAME IS = Billy SO HW? <CALL> de GM6DX +K <RX>

You will notice the plus sign here is used to generate the AR, together with no space, not all software uses the plus sign like this so you will need to check the software manual.

Likewise the equals sign sends BK, with no space, meaning break. You can see that if this was sent using PSK, although someone would understand it the format is different too short and best suited to CW.

PSK, RTTY etc;

<TX> <CALL> de GM6DX

GD OM

RSQ: 599

Name: Billy

QTH: Airdrie, Scotland

Locator: IO75

Rig: FTDX10

Ant: Moxon

Using Mix W 3.15, Windows 10

So hw cpy <CALL> de GM6DX KN <RX>

The above is a typical reply to modes such as PSK. As you can imagine even sending the ":" would be odd in a CW contact and the other CW operator would get pi$$ed off.

This is why it is better to have a few macros set for each mode type used.

73 Final CW;

<TX> <CALL> de GM6DX FB TNX FER NCE QSO BEST 73 ES GD DX TU <CALL> de GM6DX > <RX>

Like many abbreviations they are documented online, ES means "and". You will notice the > (greater than symbol) after my call, this is to create SK, together with no space again this is software dependant.

73 Final PSK, RTTY etc;

<TX> <CALL> de GM6DX

FB dear <NAME> nice to meet you today in this QSO. I wish you best 73 and good dx from bonnie Scotland, I hope to meet you agn OM.

<CALL> de GM6DX 73 TO <RX>

This gives you a general idea of what you can say or use within macros. It is important to realise that these are effectively what you would say for an exchange as you would in single side band. It is just that you will have the same typed response for each station.

When it comes to contests the exchange macros are very specific and most software that is used allows macros to be imported. This means that pre-typed specific macros can be inserted for use during the contest.

A simple search online will provide you with a link to various websites which contain these for your particular software.

Some macros are built into the software such as those used in FT8 and have little editing ability. This is because the messages are sent in sequenced manner and the digital modes being used are character length limited to fit within the timed transmission and receive periods.

To summarise macros are there to help you standardise a QSO and provide you the ability to reply in modes which are sent faster than what you can type. They ensure standardisation for particular operation such as synchronised modes or during contest exchanges.

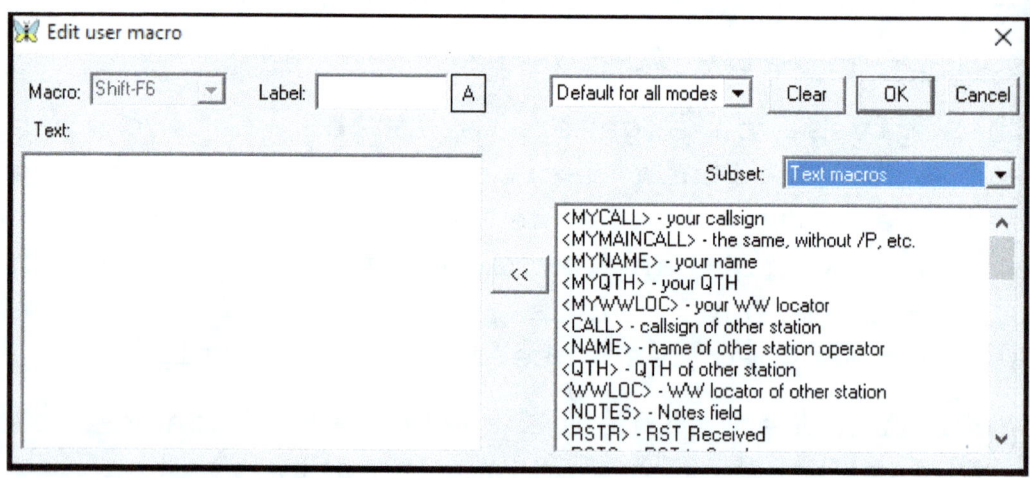

The Macro editing box in Mix W software

The Software

Chapter 7

In this chapter I am going to discuss some of the most popular software currently available for amateur radio data modes. For a more in-depth coverage of the software mentioned within this book then please visit the relevant software website for the manual or user guide. This chapter is an overview and discussion of the key features available and is not a replacement manual. Many software packages allow you to do various modes however there are a few standalone packages that are great and work well, let's look at these first.

MMSSTV – Is a free software package designed for the digital mode of SSTV (slow Scan Television) by JE3HHT, Makoto Mori. It allows you to send images and save images that you receive. Templates of transmission (or macros) are created in the software and you can use them no matter what image you decide to use. This gives you the ability to have images prepared without embedded text. The largest image size is 800 x 616 pixels in the mode of PD290 (other sizes are used). It currently can be directly downloaded on the *hamsoft* website at; https://hamsoft.ca/pages/mmsstv.php. The website gives instruction on download and install so please make sure you follow this. Once you have downloaded the software run it to see a screen similar to the below image;

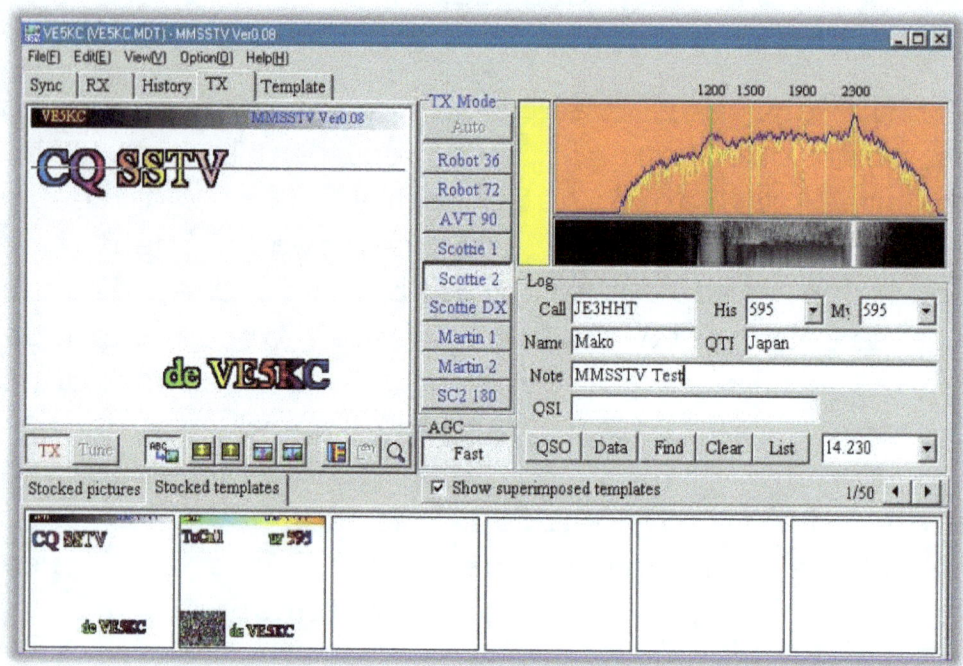

The main screen of MMSSTV software

Looking at the main screen you can see various control options such as *File* and *Edit*. Like most software if you click on these a menu will drop down giving you more options to choose from. Below these are tabs, *Sync, RX, History, TX* and *Template*. Click on these tabs and it will change the window to the left with options relating to that tab. For example click on the History tab and it will display the last image that you decoded. Click on the TX tab and it will show what your TX image, with the text template overlaying, will look like.

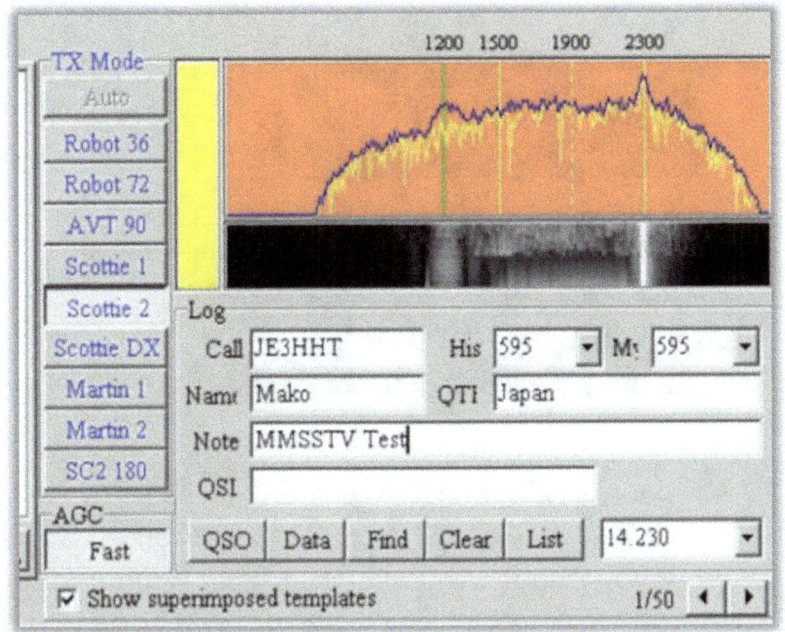

Detail section of the log and decode scope

The image above shows the details of the log. The other station's call sign, signal report sent and received, the operators name and location. Even if you are not using the built-in log for logging your contacts you will need to insert the call sign and signal report for each contact. The text template takes and uses this information for every contact and exchange. Just above the call sign field you will see a signal on the receive band scope and to the left are some of the most common SSTV modes that you can select for transmission (see manual for access to all).

Below the log section you will see lots of white boxes. This is where you will add your images for transmission.

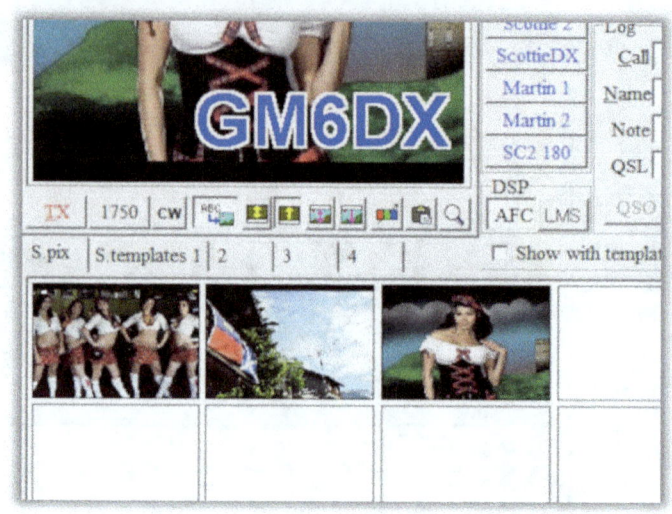

White boxes allow images to be added on MMSSTV

To add an image make sure you are on the *S.pix* tab and right click on an empty box. Next click *Load from file*. A new window appears then select the file from your PC. Once selected you will get a photo editor window where you will need to crop (reduce size) the image as well as various other editing options. Once complete your images will appear in the boxes. It is a good idea to have at least an image for a CQ call, reply and a 73 good bye message just to get you started. These windows are effectively your photo gallery allowing 100's of photos to be used. Beside the *S.pix* tab you'll notice another tab, *S.Template,* click on this and you will see various pre-programmed templates sitting in the white boxes. Similar to the images these are templates that you can edit for use in your transmissions. To edit any of these Templates simply click one of the

templates from the gallery (the white boxes).
Once the template has been highlighted by
clicking then select the *Template* tab at the
top of the TV Screen (where the pictures
appear). Now on the TV screen double click on
any of the text where a yellow highlighted box
will appear along with a window editing tool.
Here you can change colours, font and lots of
other things.

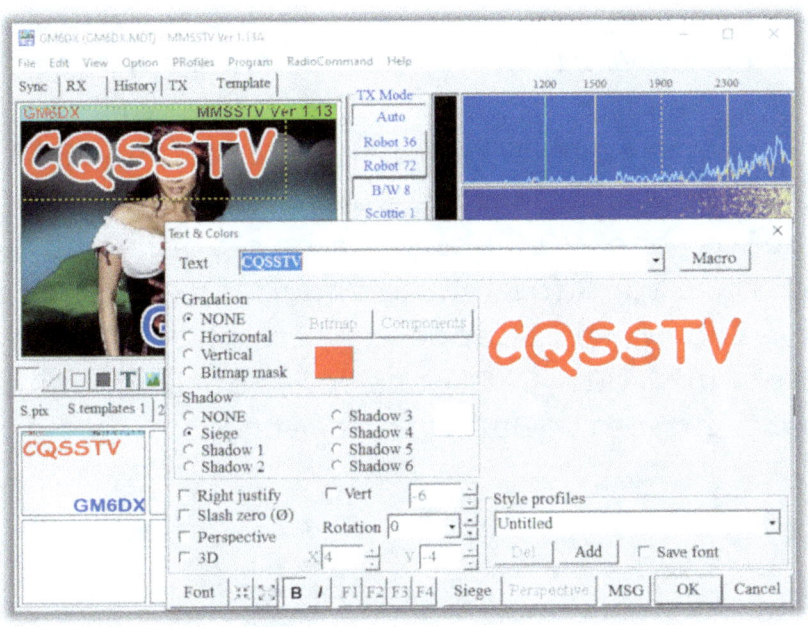

The editor window showing as well as the highlighted text

Once you have edited right click on the TV
screen and select *Save to file*. Name the
template whatever you want (new CQ for
example). Now right click on the template
gallery (white boxes as mentioned before) and
select *Load from file*. Open up your newly saved
template and it will appear in the gallery for
use. Repeat the process for all templates.

We are now ready to set up the software for use with your transceiver. Select from the top tab, *Option > setup MMSSTV(o)*. This will bring up the setup window for the software. On the *RX* tab make sure the following have a dot or a tick (selected); *Decode FSKID, Auto resync, Auto restart, Broad* and *Hilbert T.F.*

On the *TX* tab make sure the following are selected, loop back *OFF, Encode FSKID. CWID CW, TX BPF, Vari SSTV* and *Standard*, also make sure you enter your call sign in the box. Where it states PTT port select the COM port number which will make your transceiver go into PTT. Below the COM port number you will see an option called *Radio Command* click on this and a new window will appear. In this window you will set up your CAT control settings such as, COM port number, Baud rate etc.

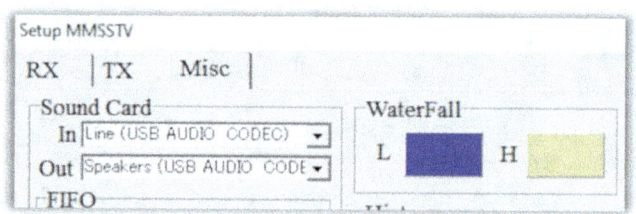

The setup window showing the various tabs

Now click on the *MISC* tab. Make sure the following are selected, *Higher, Mono, save window location*, priority of MMSSTV *higher*. At the top left you will see the sound card *In* and *Out*, here you select the sound card that is either your interface or the one which is built into the transceiver.

We have the software all setup correctly now we are ready to have an HF contact. To get started double left click on the image you want to use from your gallery. Next click the *S.Templates* tab and double click on the CQ template you created.

The TV screen will now display your image with the template overlaying the image. Select on the SSTV mode you would like to use (read the manual for full details of what each mode does) lets select *Scottie 2* for this example. Almost ready to transmit, just select the *TX* tab at the top of the TV Screen then just below your picture to the left you will see a red TX button double click this and your transmission will start.

The TV screen on the TX tab with the red TX button to the left

The transmission will start and a horizontal line will appear on the TV screen showing the transmission status.

We now wait for a reply. You will hear the other station transmissions coming through the receiver's speaker with various audio tones just as you would with an SSB signal. Select the *RX* tab above the TV Screen and watch for the decoded image to appear. It might take a few goes to actually receive something which resembles a picture so do not worry about that. Also make sure on the centre column under *RX Mode* (where the modes are listed) that *Auto* is selected that way the software will jump to the mode which the other station is using. Not all stations will reply in the same mode in which you send your CQ image. For example we transmitted on *Scottie 2* they could reply on *Martin 1.*

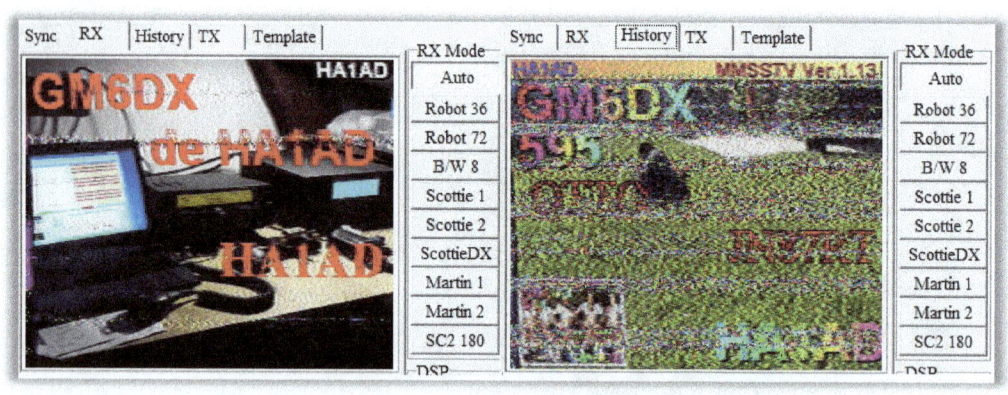

Received pictures of HA1AD transmissions during a QSO

You can adjust the soundcard volume in settings by selecting *Option > Soundcard Input Level.* If you are overloading the software with signals reduce the *volume in* settings. If you are overshooting your ALC levels (over driving) then reduce the *volume out* settings.

Once we decode an image enter the call sign into the *Call field* of the log and also enter the signal report into the *His* field. When it comes to SSTV signal reports differ from normal exchanges, look at the below table;

	R - |Readability	S – Sig Strength	P - Picture
1	Unreadable	Faint Signal	Barely perceptible
2	Barely readable, occasional	Very Weak	Poor
3	Readable with considerable difficulty	Weak	Fair
4	Readable with nearly no difficulty	Fair	Good
5	Perfectly readable	Fairly Good	Excellent
6		Good	
7		Moderately Strong	
8		Strong	
9		Very Strong	

Table showing R S P

Here are images with reference to P quality.

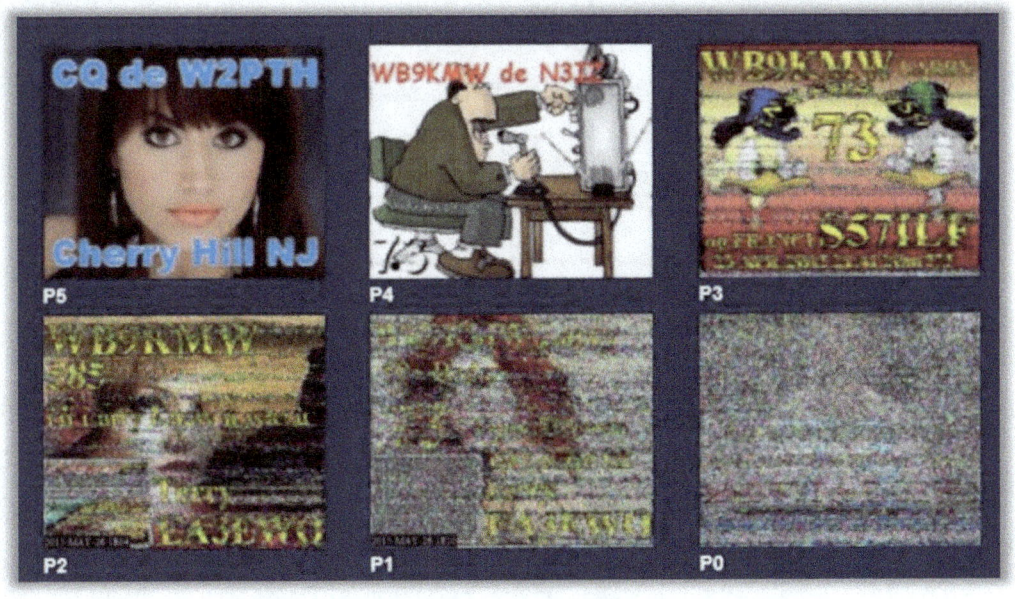

Received images with P0 (bottom right) to P5 (top left)

Now we have entered the call sign and signal report into the log we can now continue with the QSO. So double click our new image from the gallery (if we want one), go onto the *S.Templates* tab and double click the reply template. This will automatically then enter the other stations signal report and call sign into the template. It will also add the last image that you received.

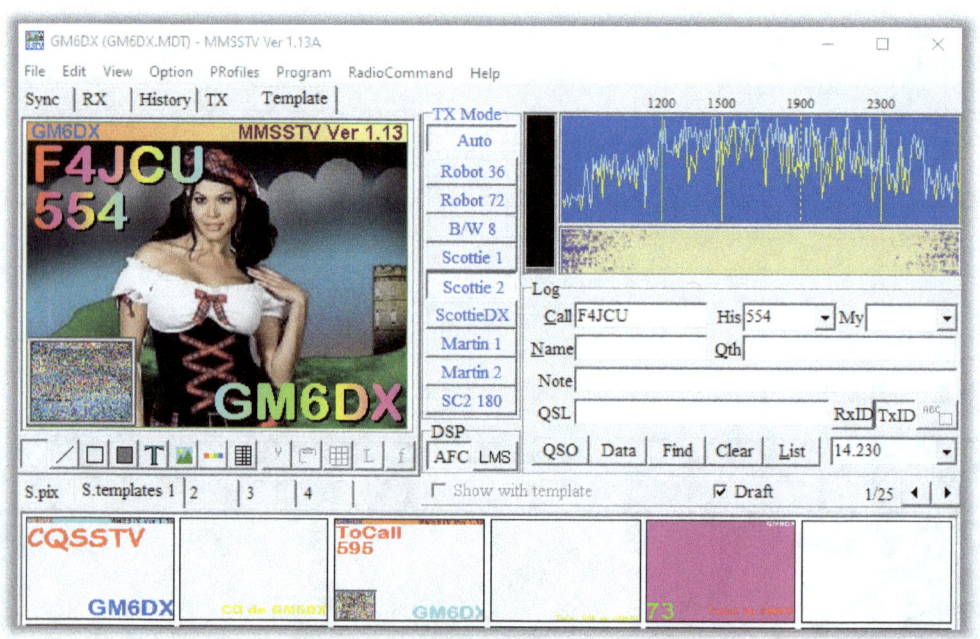

You can see the received image added to the reply, as well as signal report and call sign.

Again, we click on the *TX* tab and then the *TX* button to start our new transmission. This is the process for any new transmission wither it is a CQ call or a signal exchange. Once we transmit we wait for the next image to come through. It will be a signal report from the same station (hopefully). Once that is received we finalise our QSO with a new image, then by selecting the 73 or final template and then transmit. If we want to save any of our received images then click the *History* tab above the TV Screen, right click on the received image and then select *Save to file*. Enter a file name and location on your PC and then click *save*. No matter which software we are using the process of the QSO is the same.

Every software will have its own place for templates and transceiver setup. However the basic idea of operation is as documented in this book. MMSSTV software although free, is one the best software options for SSTV available. It is used by many worldwide and if I were to get started in SSTV today I would be using this. You can also get MMSSTV YONIQ which is a Spanish version. It is the same layout but also provides additional modes. To see how to use this software from start up to having a QSO then visit my Youtube channel at:

https://www.youtube.com/c/GM6DX

GM6DX Youtube channel banner and logo

MMTTY - Continuing along the lines of free software another great application is MMTTY. This software was also created by JE3HHT and is similar to MMSSTV but rather than being designed for slow scan television it is designed for RTTY mode. Available to download at;

https://hamsoft.ca/pages/mmtty.php

Once downloaded run the program where you will be promoted to insert your call sign. Setup is similar as most applications. We need to configure for CAT control with PTT and also for modulation (soundcard).

First we need to select the soundcard that you want to use either from an external interface or built-in soundcard. Click *Options > Setup MMTTY (o)* then select the *soundcard tab*. Once in the soundcard settings ensure you select the correct soundcard this will be done with a black dot at the relevant soundcard name.

Setup MMTTY Ver1.70K							×
Demodulator	AFC/ATC/PLL	Decode	TX	Font/Window	Misc	SoundCard	

Reception
- ⦿ Line (USB AUDIO CODEC)
- ○ Line 1 (Virtual Audio Cable)
- ○
- ○
- ○
- ○

Transmission
- ○ Speakers (Realtek High Definiti
- ⦿ Speakers (USB AUDIO CODEC)
- ○ Line 1 (Virtual Audio Cable)
- ○
- ○
- ○

MMTTY soundcard setting tab

74

Next click on the *Misc* Tab. In this tab at the top left you will see the soundcard TX and RX volume levels. Adjust these to suit your needs. Remember not to have the volume too loud that it bursts your ALC levels or overloads the software for decode. Also in the window you select the TX port wither it is AFSK (sound is selected) or FSK (Sound + Com-txd).

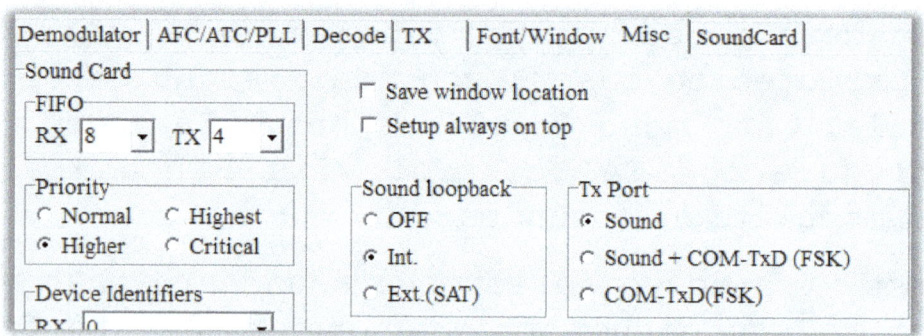

MISC tab with sound card volume and TX port settings

In the *Font/Window* tab you change the Font type of the software along with size and colour of the receive band scope, TX window and RX window.

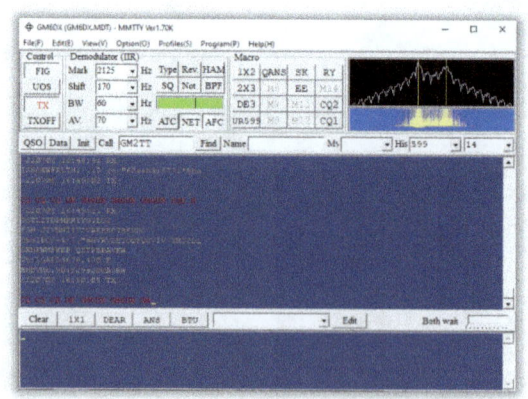

MMTTY with custom colours

Continuing in this settings box select the *TX* tab. To the right you will see a *PTT/FSK* port simply select the Com Port number that you want to use for PTT. If you are using CAT for PTT then below the PTT/FSK button you will notice the *Radio Command* button. Click on this to set, DTR/RTS as PTT, the Com Port number, baud rate and bit stops for CAT control.

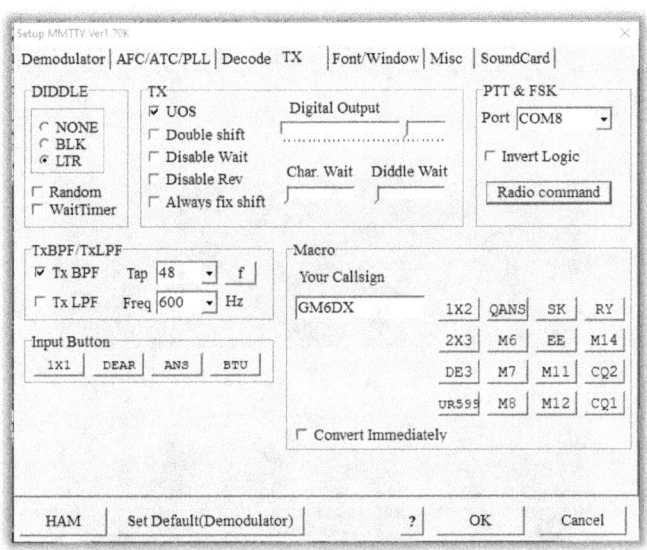

The TX Tab in settings

Radio Command settings for CAT

Now go back to the main screen and you will see a box titled Macros. Right click any macro button to edit it. MMTTY doesn't use the standard templates of use for macros already discussed in this book. This software uses symbols, example;

*CQ CQ CQ DE %m %m %m PSE K_*

Which would be the following for other software

<TX> CQ CQ CQ de <MYCALL> <MYCALL> <MYCALL> PSE K <RX>

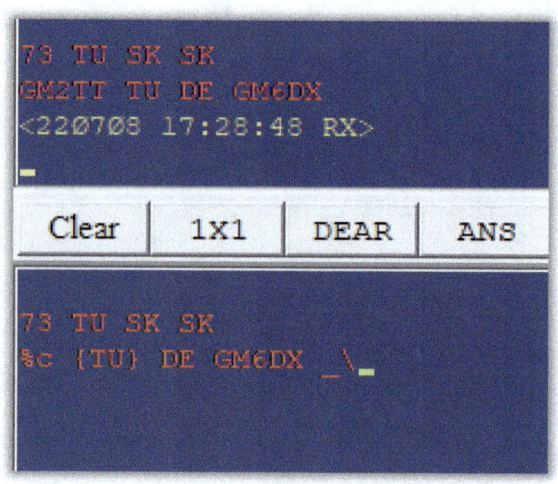

The TX window (lower) shows the actual macro message including symbols where the RX window (upper) shows what is actually being sent on air.

Once you right click on a macro you simply edit them to what you want. To send using the macro simply left click on the macro button.

Click on *Option > Setup Logging.* Here you can set up the log, including for contests, to suit your needs.

A few other settings;

Click on *View > XY scope*
Click on *View > FFT Display*

This will ensure both receive scopes are viewable on the software. If you are looking to change the RTTY shift width then on the main screen under *Demodulator (FFT)* simply change the drop down menus.

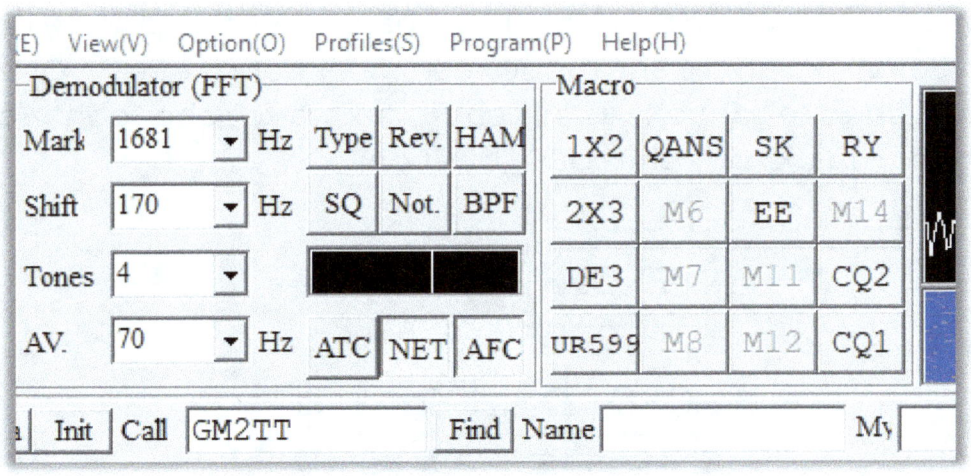

Showing the demodulator settings and macro buttons

The RTTY QSO process is simple. You first start off with the CQ macro button. Click on this a few times until you see a station replying. Their signal should appear on the XY scope as well as the receive scope on the upper right of the program. Add the call sign and other information such as name into the log bar as you go. If you don't have a call typed in the log you cannot reply with a macro after the initial CQ macro.

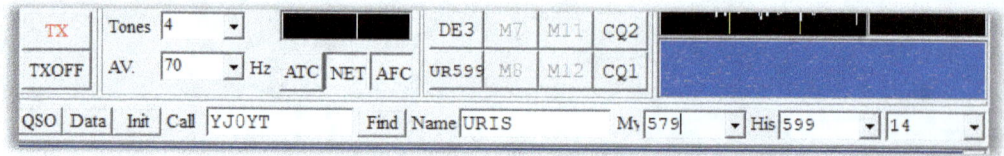

Log bar below the macro buttons

Work your way through the macros from CQ to a reply and then to your final 73 macro.

To add full details into the log click on the *DATA* button of the log bar. A window will appear with other boxes to complete. Once finished and you are ready to save the contact then click on the QSO button.

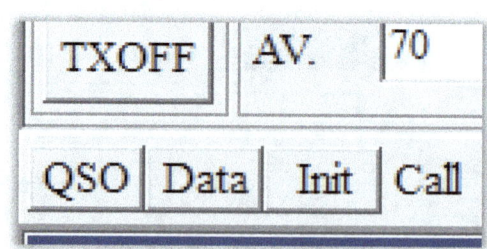

Data and QSO button on the log bar of the program

The process then starts again. You can use MMTTY software via third party software such as *Wintest*. When it comes to contests logging programs are specific and not all decoding software are suitable for contests. After all it is designed to decoded and interface with the operator not to log or run contests. As MMTTY has been around for many years and the way in which it was created logging software developers took this into consideration and designed a way to interface MMTTY with logging programs. This can take a bit of effort to get set up correctly but a short search online while return lots of "how to" guide videos showing you the process.

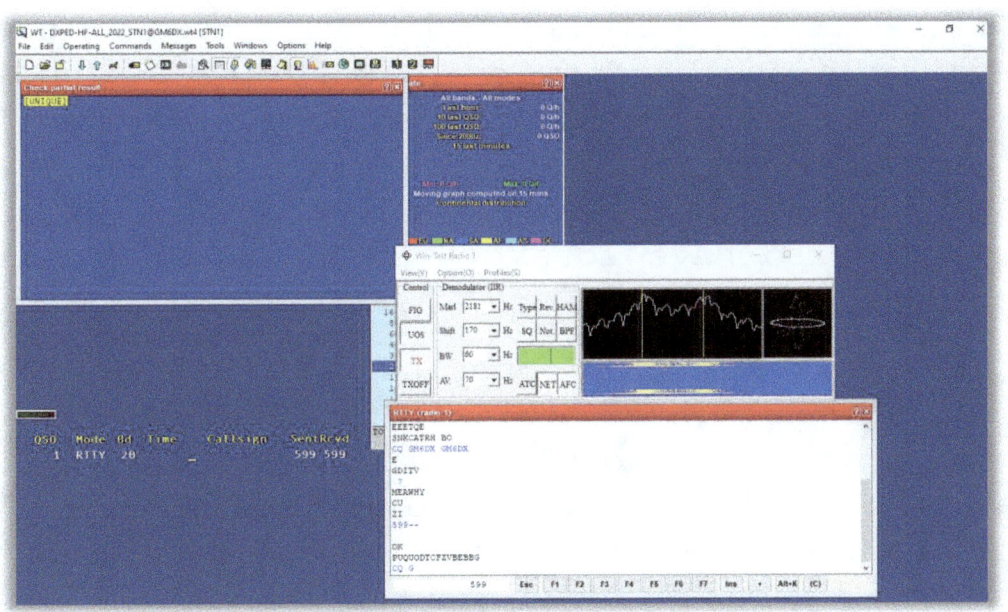

Wintest interfaced with MMTTY

In short this software is a great free program for doing RTTY. It can be interfaced to various logging programs and is very similar layout to MMSSTV so transitioning between the both should be an easy task.

Mix W – The start of this software was developed in 1992. Since then several persons have been involved in the software's current state. There are few versions available, personally I prefer the older version 3.2 however you can purchase the newer version 4.0 (or later) all this is available from the website currently at

http://mixw.net/index.php?j=downloads

This software offers almost every data mode available and the newer 4.0 software includes the newer data modes such as FT8 etc. It is of note that the software has lots of features and a study of the manual is recommended before use. I also have lots of Mix W how to videos on my Youtube channel.

Let's look at V3.2, once installed and on the main screen we need to insert our *Name* and *call* sign into the *personal details field*. This will allow the software to become registered as well as provide our Name and call sign for the macro buttons in later use.

Logging > personal data

Insert, Name (first) Call sign (as when purchased), QTH and 6 gird locator.

Next is to select the soundcard that we want to use.

Hardware > sound device settings

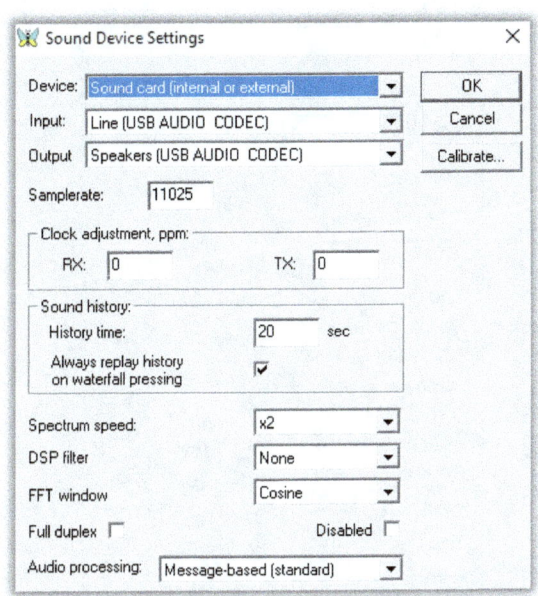

Sound Device Settings box

The same principle in selecting the soundcard and volume levels apply to all software. Make sure you select the one which is either built into the rig or as an external interface. Where it states *Device*, make sure soundcard is selected in the box.

Now that we have the sound card selected we move into the CAT settings.

Hardware > Cat settings

When the box appears you will see at the top we select the transceiver type at CAT with the model number below that. We then make sure PTT via CAT command and CW via CAT is selected (ticked). If you want to do AFSK make sure that box is also selected. The default digital mode is USB (for AFSK) also set the CW tone pitch.

82

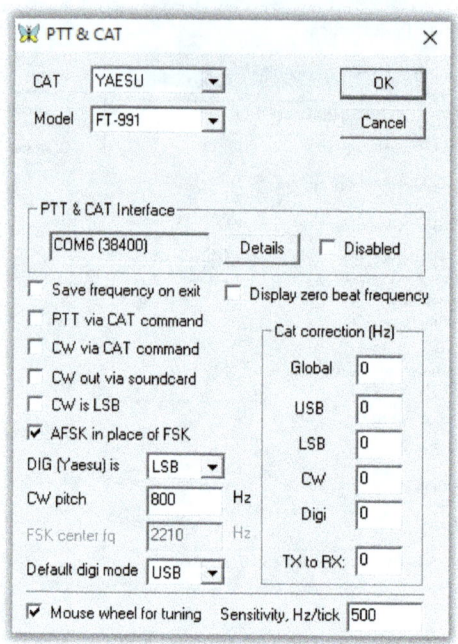

PTT & CAT settings box

We then need to set the com port settings. Click on *details* and a new box appears. This box has the com port number, baud rate, bits etc. At DTR make sure CW is selected and at RTS make sure PTT is selected. That's the CAT and PTT all set up.

Serial port settings box

In order to help you when using this software please make sure the following have a tick placed at them in the respective menus.

Logging > Merge macros

Logging > Clear QSO on new call

Logging > Use default RST

Logging > Auto search in log file

Logging > Use WAE country list

Text > Underline sent text

Waterfall > Show waterfall

Waterfall > Scroll lock

Waterfall > Show bookmarks

View > Macro bar

View > Log bar > View log bar

View > Log bar > Normal layout

Once the above have all been ticked then we are looking at the macro bar and the small wheel (like the settings button on your phone). Left click and select *Extended Bar*.

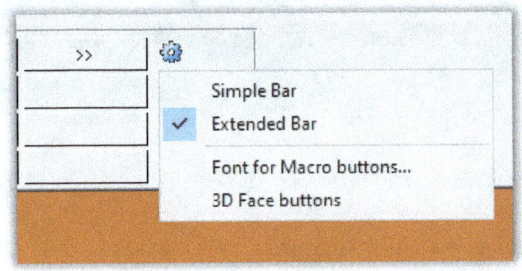

Macro settings button clicked

To edit the macro buttons, including the colours, simply right click on the macro button and fill edit text to suit your requirements.

Next, make sure you can see the waterfall at the bottom, the transmit window and the receive window at the top. If you don't then click on the grey horizontal bar and drag it up until you see two horizontal bars.

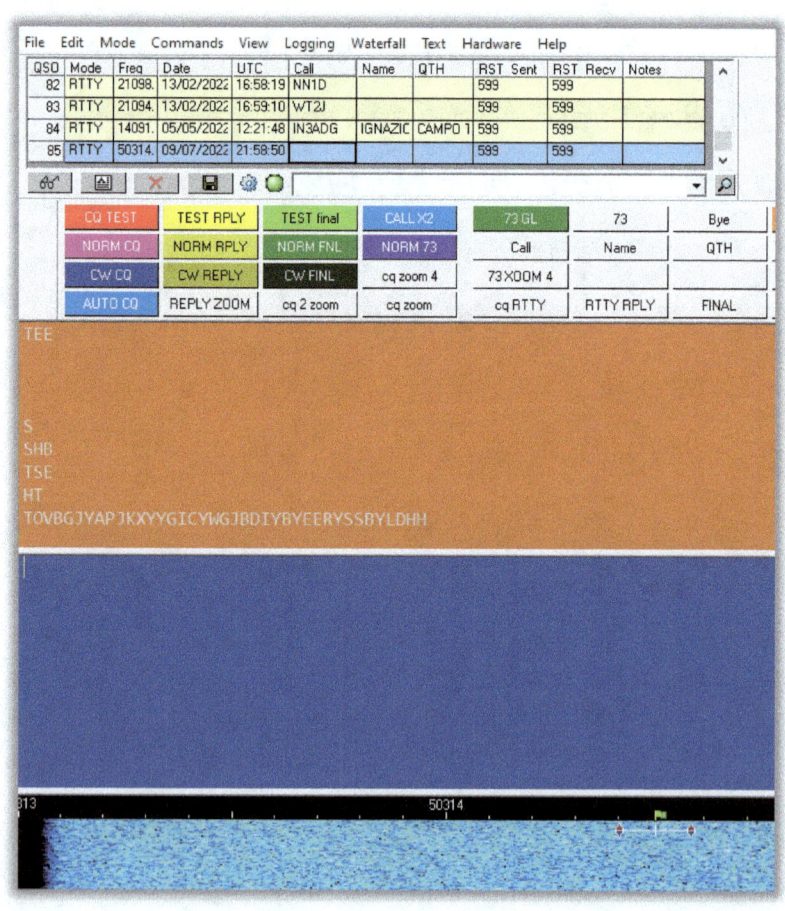

Here you can see the log bar at the top, macro buttons below it followed by the RX window, TX window and waterfall. Horizontal grey bars separate each windows. These can be moved / re-sized by clicking and dragging.

To change colours to the RX and TX windows as well as text then go to;

Text > Text colours

Change at your will. To change font type then;

Text > Fonts

Again, I have videos on my Youtube channel showing you how to edit macros and colours. This so far has covered all the flowery parts of the software now onto the modes and working stations. To select a mode that you would like use then go to;

Mode > RTTY

Or

Mode > and tick which ever mode you want. You will notice that Mix W 3.2 has 20 different modes to choose from. Depending which mode you select will depend what additional settings boxes appear. For example if you select SSTV, you will get the TV window similar to MMSSTV appearing.

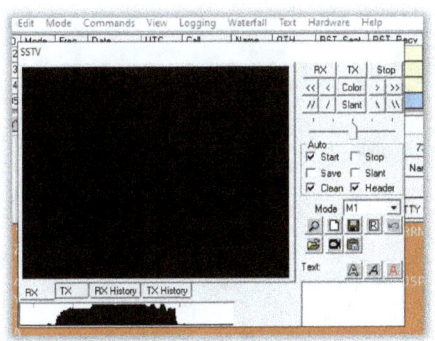

SSTV window showing in Mix W

In any mode you can access further settings to that mode by going to;

Mode > Mode Settings

Once clicked it will produce a settings box which you can edit. For example in RTTY you can change baud rate and shift size etc.

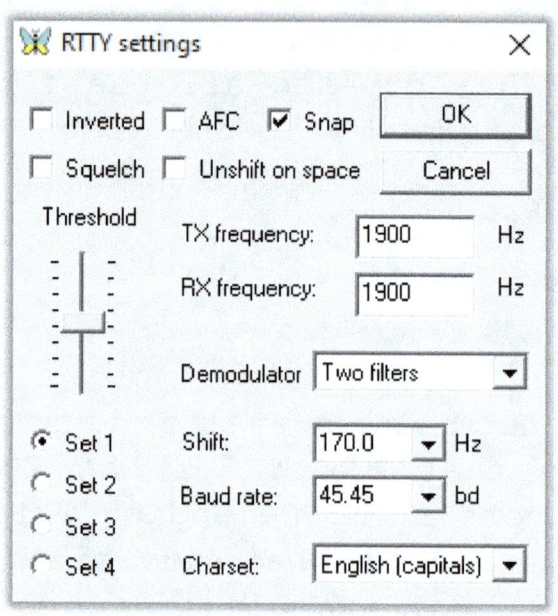

Mode settings box for RTTY

Take BPSK as another example, to change between BPSK 31 and 63 then this is done via the

Mode > Mode Settings option.

So we have selected our mode now it is time to start transmitting. If you want to use a macro then simply click on an edited one and the transceiver will go into transmit. Your message will then be transmitted.

To use manual typing then first you need to put the rig into transmit so select;

Commands > TX

The transceiver is now in transmit so simply click in the transmit window and type away. Once your message has been typed you need to put the transceiver back into receive so;

Commands > RX

You will then see a station's call sign on the screen, double click it and the call sign will appear in the log. This means we can use our macros.

The newly decode call is inserted into the log – blue line

Follow the same process as before. Going through the various macros from CQ call to 73 good bye. Once done you can save the contact into the log by clicking on the floppy disc icon to the right of the red cross below the log. You can add an auto save QSO macro option to your 73 macro this way the contact is logged automatically. This is handy as if you are like me you can forget to save the QSO before clicking on your next call sign.

Mix W also offers macros to be imported as well as contest logs and contest statistics. It also offers spotting networks to *dxsummit.fi*.

This software is one of my favourites for doing all modes (except FT8/FT4). It has so much to offer and very simple to use once you have it set up. Be sure to check out my Youtube channel for various videos on Mix W.

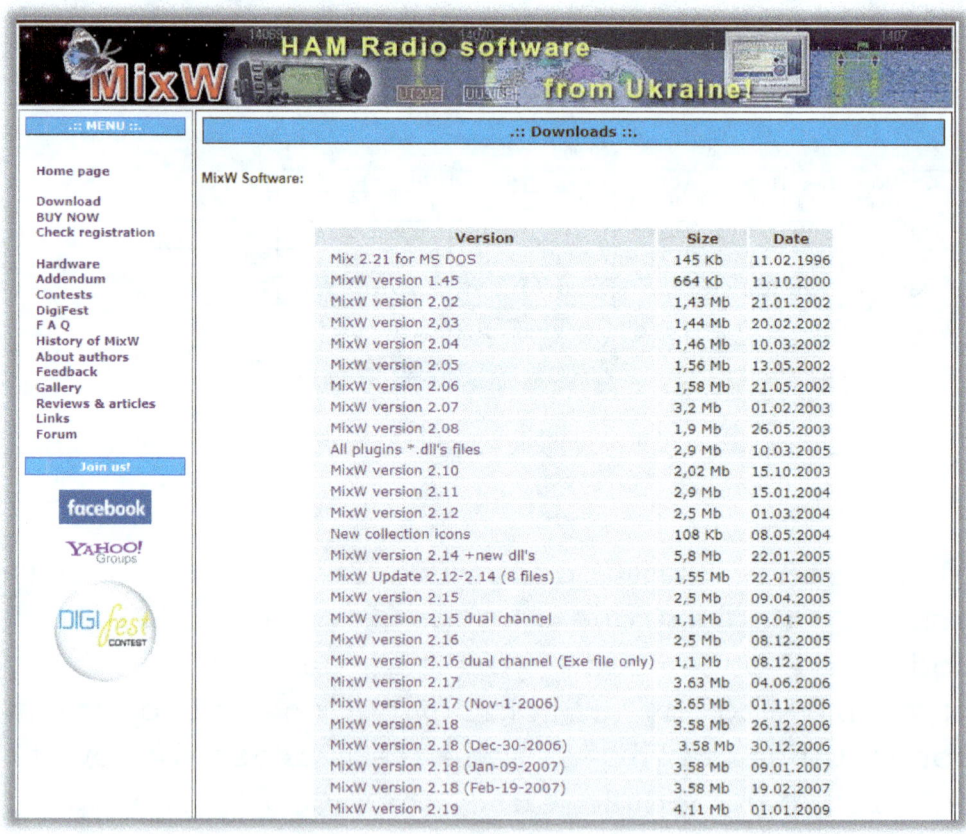

Get various versions of Mix W from MixW.Net

MRP40 – When it comes to CW one of the best programs around is a paid product known as MRP40. You can purchase the software from the below website;

http://www.polar-electric.com/Morse/MRP40-EN/

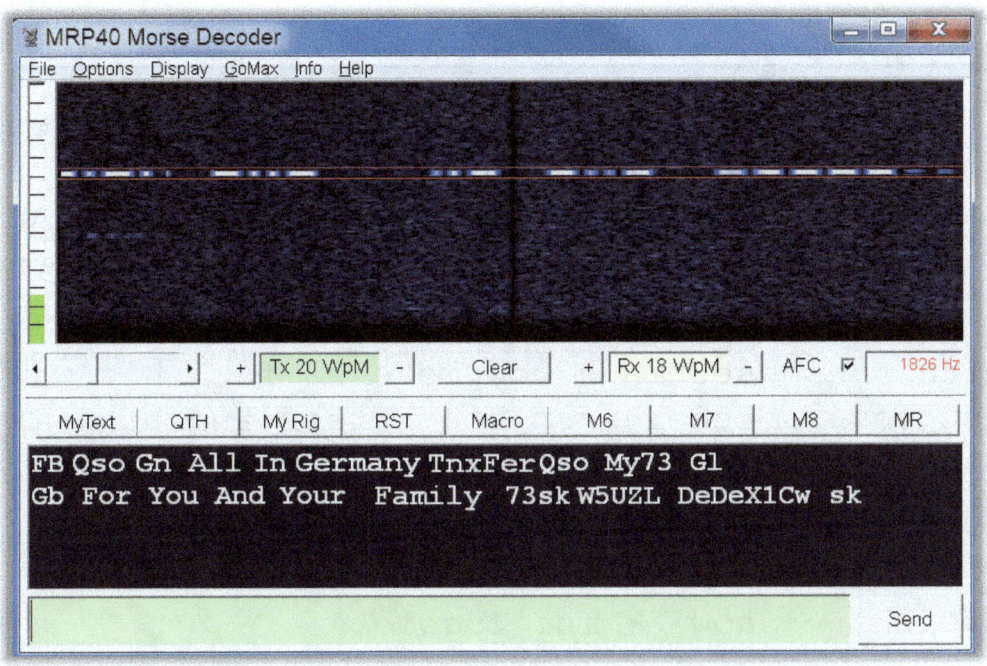

MRP40 Software on the main screen.

This software allows decoding of CW as well as the ability to send CW by simply typing into the transmit bar (green on above image).

Once you have downloaded and installed the product there are a few settings that we need to go through in order to get you set up to use the software. First is to make sure the following are selected;

Options > RX Settings > Weak Signal Decoding

Options > RX Settings > CW Filter Wide (select narrow for contests etc)

Options > show > soundcard then select the soundcard that you will be using for Release RX.

Options > TX Settings > Send Via Com Port

A new box will appear. Enter the correct com port number for your interface and make sure that DTR is set as the *send* pin. Untick any disable PTT function options. Enter save and close the box down.

With these settings complete so far this will get you to receive on the screen and transmit when you type in the *send box*. Now that we have the basics done let's look at some of the other features.

Options > Show > Mini logbook

This will bring up a window which has basic details for you to log the QSO. To the right of this window you will see *open*. When clicked it will open up the log in text format. Beside the open button is *save*, click this to save the details of the QSO into the log.

Options > Show > TX macro box

This will bring a small box up where we fill out our call sign, name, QTH, signal report as well as a serial number exchange for contests, should it be needed. The main macros are edited on the main window screen below the RX window (top window). Simply right click on the macro you want to edit, change the text and once you're happy select save and close. You can type directly into the macros without the need for any commands similar to <TX>, an example below;

CQ CQ CQ GM6DX GM6DX PSE K

When that macro is pressed it will appear in the green send window, in order to send the macro text push the *SEND* button.

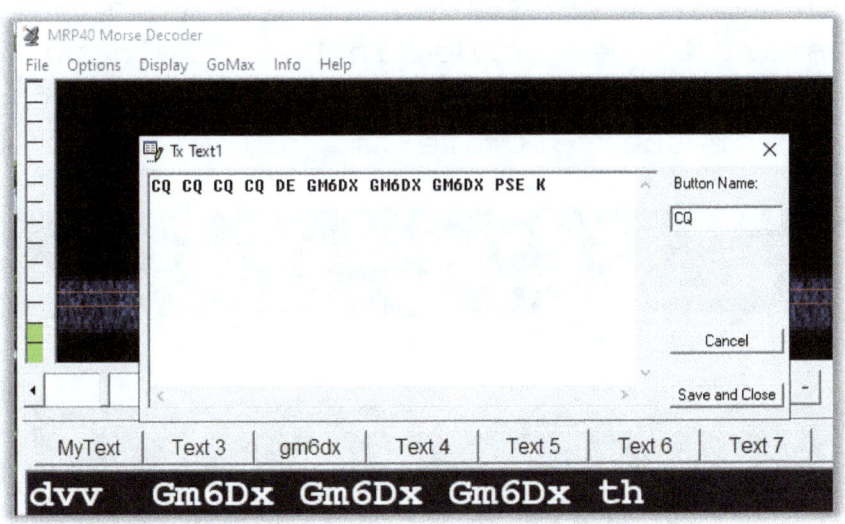

How the macro button appears under edit

Options > Show > Text font size

Here you can edit the RX and TX window size of text, font and colours.

To use the software turn AGC off on your transceiver. Make sure your rig is set up for

DTR CW keying if you are using the USB CAT control. You may also need a wide filter on the rig such as 3kHz but start off with 500Hz (narrow filter).

To the upper left of the software main window you will see a green volume bar. When it remains no lower than 10% of the bar length, then that is a good indication that the audio line is at a good volume for the software. If not adjust until it is within the levels.

Once volume has been adjusted you will start to see dots and dashes coming from the left hand side of the screen to the right and CW will start to be decoded in the RX window. Once you see these appear simply click on the CW signal for the software to follow.

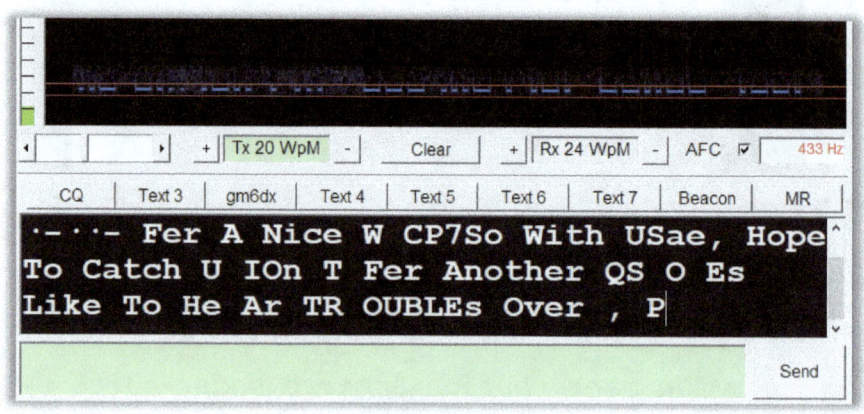

The CW being decoded

Above the macro buttons you will see RX words per minute this is the speed of the station you are listing to. You can adjust the TX words per minute to match that of the sender which button is also above the macros.

In order for you to reply simply click a macro where the macro text will be added to the bottom TX window and then push send or hit enter. You can free type in this window also even when *send* has been pressed. So you could type the stations call sign followed by yours then press send. Thereafter as you type it will still continue to send until there no more text left in the TX box to send.

CW QSO's are the same as any but abbreviations are used. Rather than bore you to death in this book then have a look online at Morse code abbreviations you will find a list containing quite a few, here is an example CW reply;

FB OM TNX FER RPT OP = BILLY B I L L Y SO HW? OY2T DE GM6DX

Once you have made your exchange you can log it in the MRP40 log or use your own logging program. The log details are added into the mini log which you already set to be displayed.

One thing I will say about this software is that below 15 words per minute (CW speed) or in very noisy band conditions it is not great at decoding and what is displayed can be questionable. Overall this software is one of the best available it is good for you to get on air with CW and you are of course doing the first true data mode known. Why not give it a go and see how you get on.

WSJT-X - is one of the most popular free software for doing modes such as, FT8/FT4/JT9/JT65/WSPR/Q65/MSK144 and FST4. All these modes work on the same principle of synchronised timed transmission and receive. Before you proceed it is worthwhile to install and use dimension 4 software which is an accurate time keeping software for your PC's clock.

http://www.thinkman.com/dimension4/

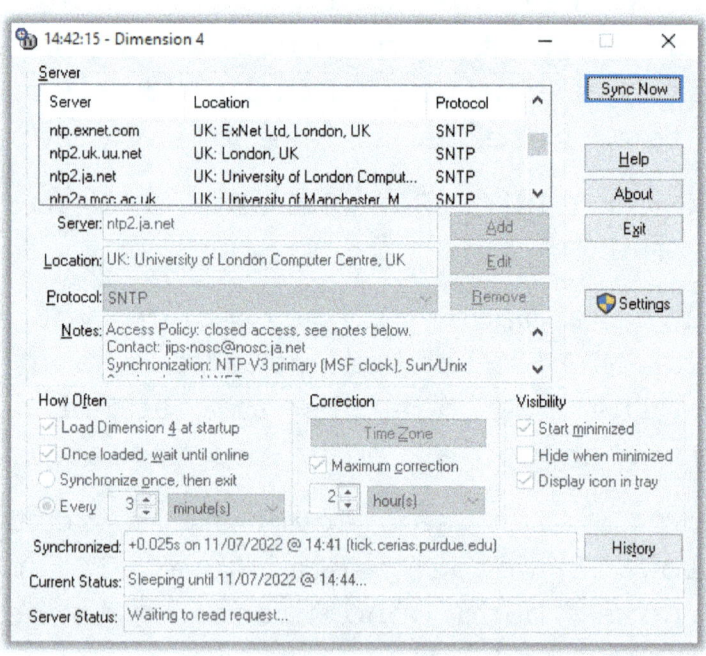

Dimension 4 software to ensure PC clock is accurate, if not it can prevent bad or no decodes.

Now visit the below website and download the latest version of WSJT-X.

https://www.physics.princeton.edu/pulsar/k1jt/wsjtx.html

Install the software and run it. If you are not promoted at initial running of the software then follow the below;

File > Settings > General

Once in general tab enter your call sign and 6 grid locator reference. Then ensure you have a ticked in the following boxes

Blank line between decoding periods

TX messages to RX frequency window

Double click on call sets TX enable

Disable TX after sending 73

Allow TX frequency changes while transmitting

Once the above has been selected click on the *Radio* tab at the top. Here will select all the settings for CAT and PTT.

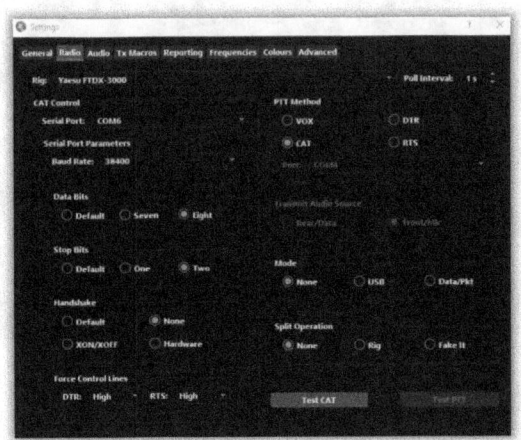

Settings box being displayed at Radio tab

In the CAT box select your serial port number for CAT as well as baud rate. Across from this at *PTT method* select CAT. At *mode* select none. At split operation select *fake it.* To finalise select the bit, stops and DTR lines to suit your rig. That is the radio interface settings complete, now select on the *Audio* tab at the top.

At *INPUT* and *OUTPUT* select your sound card. Beside these make sure both have *mono* as an option. This completes the sound card options.

The remaining settings such as colours will be down to personal choice and will require you to play about with the *Options* to see what you like, or don't like as the case may be. Save those options and go back to the main screen

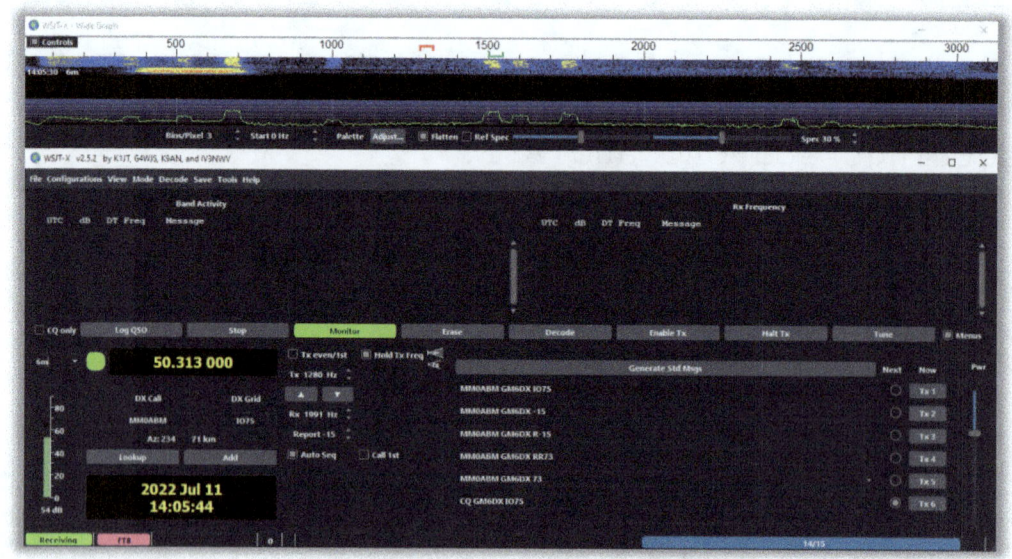

WSJT-X main screen (in dark mode) with the wide graph at the top of the actual decode and main interface window.

Now that we have all those settings done you should be decoding signals on the wide graph. If not go to mode and select one of the modes available. Depending what mode you select will depend what other interface boxes appear. Let's choose FT8. At the bottom right of the main software you will see the macros for sending.

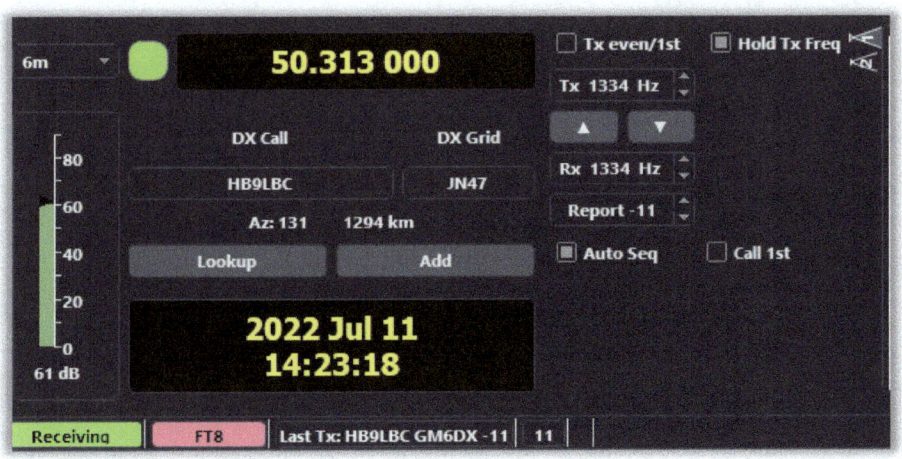

Macro buttons of the software

To the left of this you will see the volume bar (green) along with frequency and TX sequence control.

CAT / Frequency along with sequence control

If we want to TX on the even or first cycle simply tick this box. If we want to hold the same TX frequency for every station we call (recommended) then tick this box. If you want the macros to cycle through in order (recommended) then tick *Auto Seq* box. Above these options you will see a *band activity* window to the left and a *RX Frequency* window to the right. Every station that is decoded on the opposite cycle of you, than what you are transmitting on will appear in the *band activity* window.

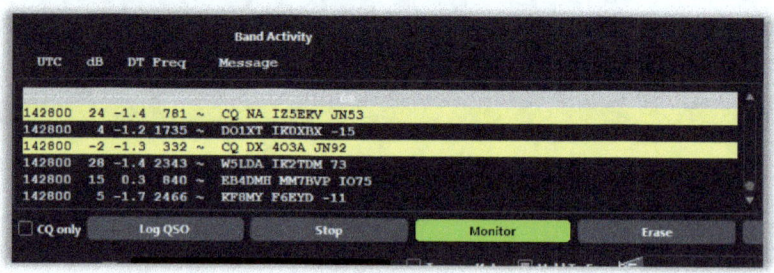

Band Activity window of the software

The call signs to the right are the call signs which we are actually receiving. In order to have a QSO with one of these station go to the wide graph and double click somewhere which has a free space (ie not a signal on it).

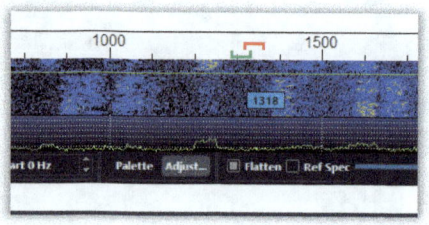

The red line is the TX frequency and the green line is the RX frequency.

Now click on the up arrow, which will jump the TX frequency to where we have just double clicked on the screen.

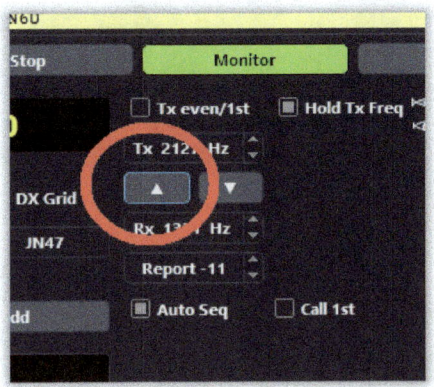

The up arrow to click on once we have double clicked on a free space on the waterfall.

That means our TX frequency is now on the space that we just ticked so click on the *Enable TX* button to start the QSO. Make sure you have selected the bottom macro which is the CQ macro.

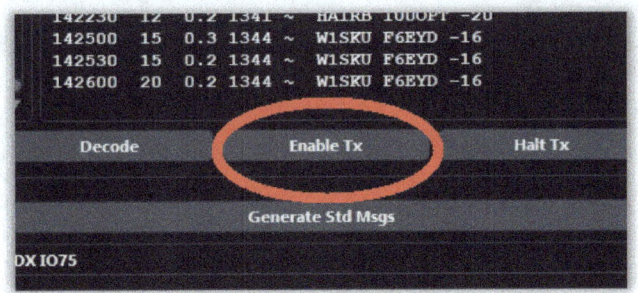

The enable TX button to start the software into TX cycle.

Once a station replies to our signal you will see a red bar appearing in our RX frequency window. That is the station calling us. Double click on the red bar to start the macro cycle.

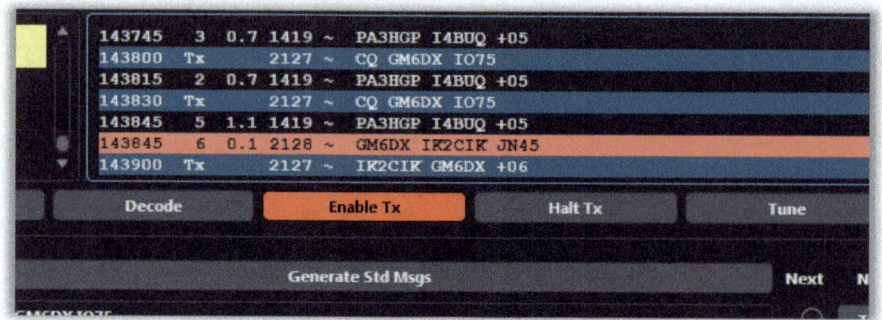

A station calling shows as a red bar

Now you can sit back with a few beers and let the software cycle through the macros to complete the QSO exchange. Once it has sent the 73 or RR73 macro, which is the final, the log window appears for you to log the QSO. I recommend using the built-in log.

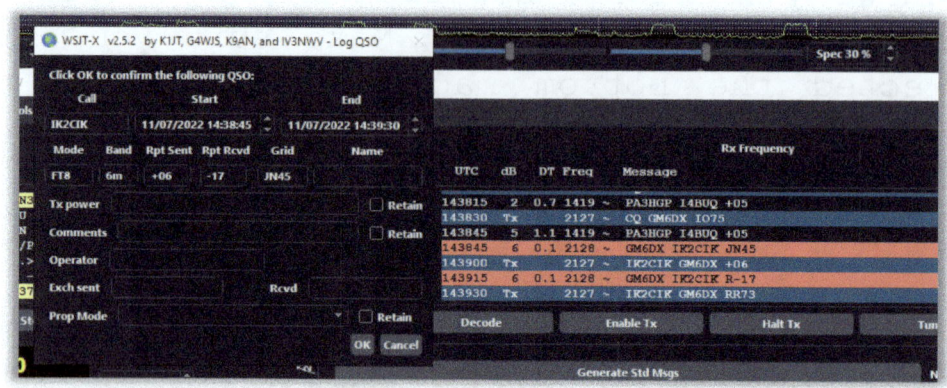

The QSO cycle complete and log bar has appeared.

To save the QSO click OK on the log window that appears. This cycle or process is the same for any of the modes selected via the *Mode* menu at the top. You can see why it is not the most human interfaced of data modes compared with some of the traditional modes such as PSK. The signal exchange reports are also in dB. A typical -05 is minus 05dB that you are being received at the other station. It takes roughly about +10dB to be equal to a CW signal on air. These signal reports are affectively below the audio range meaning that they are that weak in strength you couldn't hear them with your ears via your receiver, especially if they are below +10dB.

The software takes a bit of getting used to but you can learn how to operate it very easily and quickly. To the right hand side of the software you can see a *PWR* bar. Sliding this up or down increases or decreases the output volume of the software. This is something to keep an eye on especially with ALC levels.

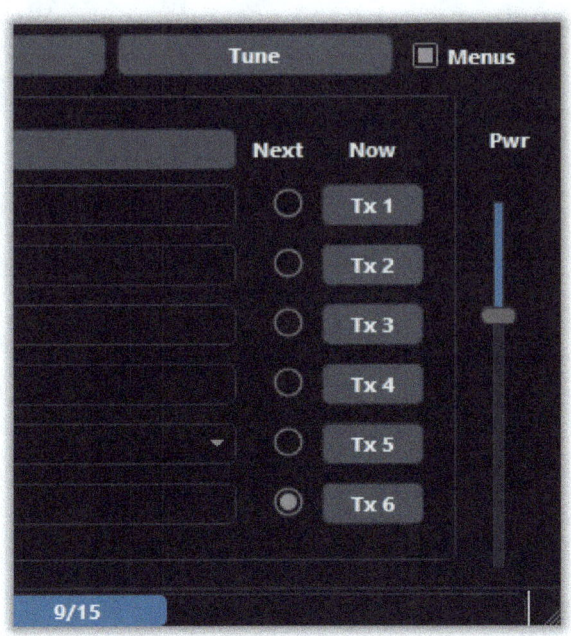

The PWR bar to on the software

When you select MSK144 from the mode menu a different waterfall appears. This shows the "ping" as the software is designed for meteor scatter or aeroplane ping.

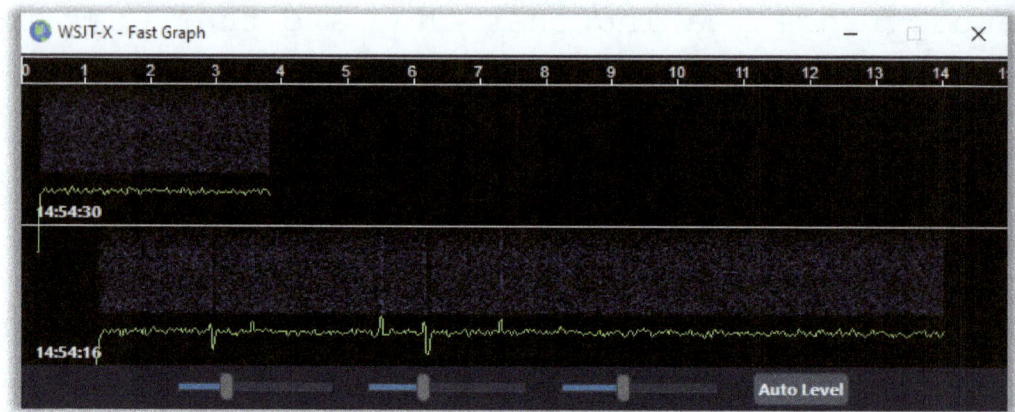

The MSK144 receive graph that appears when doing MSK 144

You will also notice that I have the WSJT-X software in dark mode. If you would like to do this then; Right click on the short to WSJT-X and select properties.

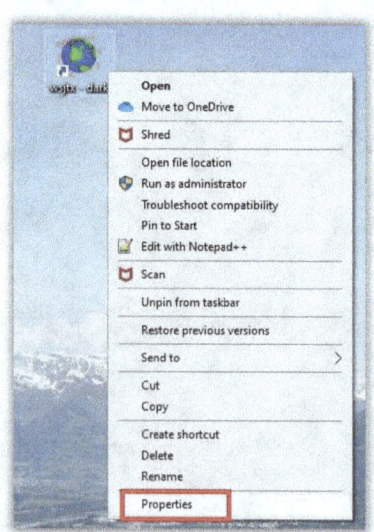

Right click on the short cut select the properties option.

Once in properties click on shortcut. At the target tab enter this *AFTER* the .exe

-stylesheet :/qdarkstyle/style.qss

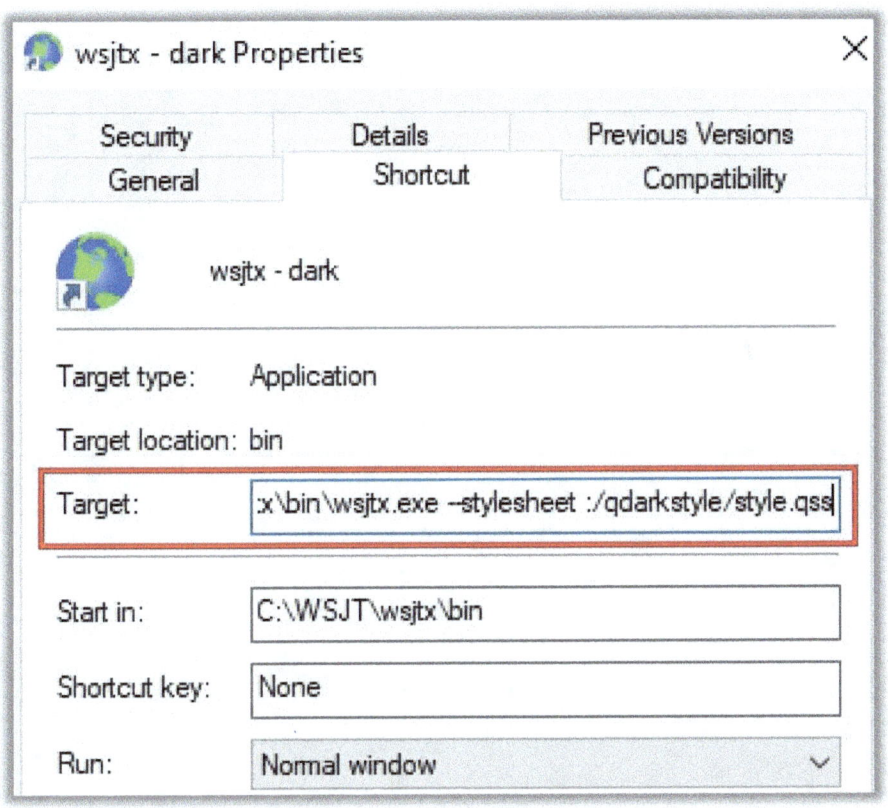

The properties field and what should be typed

Then click *OK,* then *apply.* Every time you click on that short cut it will run WSJT-X in dark mode. There are plenty of videos online regarding this software should you get stuck with it or the QSO process in general. Be sure to watch them rather than spend hours of your life that you will never get back looking for that box you have failed to tick.

The End
Chapter 8

This brings my short book to an end. Firstly I would just like to say that there are plenty of other software available other than the ones covered in this book. Some are better are doing certain things than others. However this book was designed to give you an overall idea of how data modes work and their principle, not to be a sales rep for all these various software manufacturers. If you find software out there which you prefer and can work it then happy days I am glad for you and you should stick to using it.

The whole purpose of the book is to get new people into data modes who would never have thought about giving it a go. If this book has helped you do this then I have achieved what I set out to do.

Hopefully with the information from my book you should be able to have a better understanding of the origins of data communication and the reason why we have so many available today including the background of how we got there.

This book provides you with various interface design options available for you to construct on your own. It also looked at the commercially available to purchase interfaces and what to consider when purchasing. It covers the various interfaces needed in order to actually do data mode communication such as CAT and audio.

It also covered the various modes used in today's communication. Looking at the background of them, key features and what to consider when doing each mode. The common software used by amateurs was then covered showing you the key points needed to operate and function the software correctly for various modes. This book also covered what you need to consider when setting up your own macros and the QSO process used.

In short everything has been discussed and covered for all that you need to know to get you started in data mode communication. Yes there are more in-depth books published that you can read (if you don't fall asleep first) covering various data modes for that more in-depth level of understanding.

However the content of this book covers what you really need to know. That is the basics to get you going on air and the rest will follow.

As well as this book I have numerous videos on my Youtube channel at

https://youtube.com/c/GM6DX

Which should provide you with some visual guidance of the modes and software setup discussed here.

I hope you have found this book helpful to some degree and that I can have a data mode QSO with you over the airwaves in the near future.

73, GL, GD DX

Billy GM6DX

www.ingramcontent.com/pod-product-compliance
Lightning Source LLC
Chambersburg PA
CBHW051334170526
45166CB00002B/805